Benjamin Hill

An epitome of the homoeopathic healing art

Containing the new discoveries and improvements to the present time

Benjamin Hill

An epitome of the homoeopathic healing art
Containing the new discoveries and improvements to the present time

ISBN/EAN: 9783337210700

Printed in Europe, USA, Canada, Australia, Japan

Cover: Foto ©berggeist007 / pixelio.de

More available books at **www.hansebooks.com**

AN
EPITOME

OF THE

Homœopathic Healing Art,

CONTAINING

THE NEW DISCOVERIES AND IMPROVEMENTS
TO THE PRESENT TIME;

DESIGNED

FOR THE USE OF FAMILIES AND TRAVELERS

AND AS A

POCKET COMPANION

FOR THE PHYSICIAN.

Revised Edition.

BY B. L. HILL, M.D.,

essor of General, Special, and Surgical Anatomy, Late
urgery, Obstetrics, and Diseases of Females and Children, in the
W. H. College, Author of the "Homœopathic Practice of
Surgery," &c., &c.

DETROIT, MICHIGAN:

PUBLISHED AT DR. LODGE'S HOMŒOPATHIC PHARMACY.

TABLE OF REMEDIES.

TRAVELER'S CASE.

INTRODUCTION.

This work contains in a *condensed form* a very large portion of all that is practically useful in the treatment of the diseases ordinarily occurring in this country. The symptoms are given with sufficient minuteness and detail to enable any one of ordinary capacities of observation to distinguish the complaint; and the treatment is so *plainly* laid down, that no one need make a mistake. If strictly followed, it will, in a very large proportion of cases, effect cures, even when administered by those unacquainted with the medical sciences generally. It has been written from necessity, to meet the demands of the community for a more definite work in a concise form, that should contain remedies of the most reliable character, with such directions for their use as can be followed by the *traveler on his journey*, or by families at home, when no physician is at hand. It might seem to some preposterous to speak of a *demand* for another *domestic* Homœopathic Practice, when half a score or more of such works are now extant, some having come out within a very short time. The demand arises, not from the want of books, but from the defects of those that exist. There is in most of them, too little point and definiteness in the prescriptions, and a kind of vague doubting recommendation noticeable to all, which carries the impression at once to every reader, of a want of *confidence* by the author in his own directions.

Again, in some of the works there is too much confusion, the symptoms not being laid down with sufficient clearness to indicate the best remedy. Some of the works are unnecessarily large and cumbersome, while the real amount of valuable practical matter is comparatively meagre, obliging the reader to pay for paper and binding without the contained value of his money.

This work is my own, being the result of my practical experience and observation. I have introduced several remedies that, though they are familiar to me, and have been used in my practice for many years, are, nevertheless, comparatively strange and new to most of the profession. For provings of these I refer to " New Homœopathic Provings," edited by E. M. Hale, M.D., an octavo volume of 448 pages, published by Dr. E. A. Lodge, Detroit. Their use, e directed in this work, is in strict accordance with their homœopathic relation to the symptoms for which they are prescribed.

Some may object to my practice of giving several remedies in alternation or rotation and in quick succession. To such I would say, When you try this mode of practice and on comparing it with the opposite one of giving only one remedy, and that at long intervals between the doses, find my mode to be less successful than yours, *then* it will be time for you to make your objections. *You* may rely upon the vague hypotheses of the books, and give. your high dilutions singly, at longer intervals, and let your patients die for want of *real* treatment, while I will use lower dilutions and give two or more remedies in quick succession and cure mine. I only speak what is in accordance with universal observation, where the two modes are compared on equal footing, when I affirm that, while the former *may* effect some cures, *most* of the recoveries under it are spontaneous and unaided. the latter *does* cure; the disease being arrested by the medicine, and the proportion of unfavorable terminations is much less under the latter than the former course I know many learned and successful practitioners who have substituted low dilutions and the giving of several remedies in quick succession for the old mode of high attenuations and long intervals of single remedies, all of whom still adhere to the low, while I have yet to hear of the man who has gone *back* to high single remedies and long intervals. My reason then, for the course here laid down, is, that it will *cure* with more promptness and certainty. If others are so prejudiced as not to *try it*, they will still remain in ignorance of the *best practice*, and their patients will be the sufferers.

In reference to the fear that is expressed that if one medicine is given too soon after another, it will antidote the former, I have simply to say, I have no confidence in the hypothetic antidotal powers of the medicines one over another, as laid down in the books. It has not been verified by experience, and has no foundation in truth. It is true that one medicine will remove morbid symptoms that might be produced by an overdose of another; but both being given in the ordinary medicinal doses, neither of them to such an extent as to produce sensible symptoms.

if given alone, would not, if given in quick succession, prevent each other from acting to remove their own peculiar symptoms that exist in the system at the time. So if we have the symptoms that are found in two or more different remedies present in the same attack, as is often the case, we may give these several remedies one after another, with confidence in their curative effects for the symptoms they represent.

This has been my practice, and it has been eminently successful, and therefore I commend it to others.

Dr. Lodge is responsible for the additions made in brackets [ ].

PREPARATION OF MEDICINES.

As it often becomes necessary for the practitioner to make more or less of his own dilutions and attenuations, some brief instructions especially to new beginners, may not come amiss.

Medicine is prepared by mixing it with distilled water, or 98 per cent. Alcohol, which has been specially purified for homœopathic use, or if solid and dry, by reducing it to powder and triturating (rubbing) it in a mortar with pure Sugar of Milk. The liquid· is called *dilution*, the powder *trituration*. The attenuations made at the decimal (1-10,) ratio and numbered 1, 2, 3, &c., by putting ten drops of the liquid with ninety drops of. Alcohol, or ten grains of the powder with ninety grains of Sugar of Milk for the 1st, and ten grains or drops of the 1st with ninety more of Alcohol or Sugar of Milk, as the case may be, for the 2d, and so on to any desirable extent.

Accuracy is very desirable, but the practice of *guessing* at the amount as pursued by some, is anything but accurate. When one makes his dilutions by putting the fluid into a vial and "*pouring it all out,*" *guessing* that he has a *drop*

x

left which is to medicate the ninety-nine drops of Alcohol or water, he may put in by guess, I am inclined to *guess* that he knows nothing, *accurately*, as to what dilution he is making. (See Laurie's Practice, introduction, also Jahr & Gruner's Pharmacopœia and Posology.) For if the vial is small and quite smooth there may not be a drop left, or if it is rough, there may be several drops. Have true scales for weighing solids, and a graduated measure marked from ten drops up to one hundred for liquids; then *always* weigh or measure *accurately* the medicine, as well as the substance with which it is to be attenuated. The measure and mortar, after using them for one medicine, can be cleaned preparatory for another, with scalding water, rinsing them with purified Alcohol, then drying. Never smoke or chew tobacco in any place, but if you are such a *slave* to habit, that you must do it despite your good sense and better judgment, never do either, or have tobacco or any other odoriferous substance about your person when you are preparing medicines, or they are exposed to the air. Keep the medicines excluded from the light and air as far as practicable. *Triturate* the powders thoroughly for an hour or more upon each, and shake the dilution from fifty to one hundred times, more for the higher attenu-

ations. It is better to *medicate pellets* in large
bottles, filling them half or two-thirds full, put
in just liquid enough to wet every one, but not
so as to dissolve any. Shake them until all are
equally wet, and let them stand for four or five
days, if practicable, shaking them up two or
three times a day until all are dry.

ADMINISTRATION OF REMEDIES.

THE remedies are either in the form of tinctures, dilutions, pellets or powders. The *Pellets* may be taken dry upon the tongue, allowed to dissolve and swallowed. The dose for an adult is from 4 to 7; for an infant, from birth to one year old, 1 to 3; from one to three years, 2 to 4; from three to ten years, 3 to 5 pellets; after ten, same as an adult. 15 or 20 pellets may be dissolved in a gill of water, and a tea-spoonful dose given at a time, being particular to stir it until all are perfectly dissolved, stirring it before each dose. *Powders* may be taken in the same manner, upon the tongue, a dose when dry, being about the same bulk as of the pellets as nearly as practicable. If put into water, to a gill of water add of the powder about what would lie on a three cent piece. If the dilution is used, add 5 drops to a gill of water, and use tea-spoonful doses as above directed. The length of time between the doses should be, in Dysentery and Diarrhœa, regulated by the frequency of the discharges, giving a dose as often as the evacuations occur. In acute and violent diseases, the doses should be repeated oftener than in milder cases—about once an hour as a general rule is often enough, though in some cases they should be given in half an hour or oftener. In mild cases, once in two or three hours is often enough, and in chronic cases once or twice a day.

Bathing —The surface of the body should be kept clean, as far as possible, and to this end, in summer, should be well bathed at least once a day. In winter, though useful, it is not so indispensable; still no one should neglect the bath more than a week, and all ought to bathe at least twice a week, even in winter. The bath should be of a temperature that is agreeable and the room warm, especially for a feeble person. It should be so applied as not to give a general chill. as such shocks are always hurtful.

The teeth should be kept clean and free from tartar. They should be cleaned every morning and after each meal.

· Clothing —The feet, legs, and arms should be warmly clothed, especially the *arms*, as an exposure of them to cold is liable to induce affections of the lungs, and to aggravate any existing disease of those organs. By exposure of the feet and legs to cold, diseases and derangements of the female organs, even in young girls, are induced ; and one prolific cause of female weakness is to be found in improper dressing of the feet and legs, while the *lung affections* of females, now so fearfully prevalent, are traceable in a great degree to the fashion that has prevailed for a · few years, of exposing the arms to cold. In winter and spring when the weather is mild, but there is snow, or the ground is damp, more clothes are necessary than when it is freezing hard and the air is dry.

Diet.—The diet of the sick should be nutritious, but at all times simple, free from greasy substances, and from all stimulating condiments whatsoever, as well as from vinegar, or food in which vinegar is used.

In short, let the food be nutritious, easily digested, small or moderate in quantity, and free from all "seasoning," except salt or sugar; and if salt is used at all, let the quantity be very small, much less than would be used in health.

Diarrhœa.—This disease consists in a looseness of the bowels, generally accompanied with pain in the abdomen, more or less severe. It sometimes occurs without pain, but is *then* attended with a sense of weakness, and a general feeling of uneasiness. It prevails mostly in the warm seasons. It is not usually considered a very dangerous affection, except during the prevalence of *Cholera*, or in children during hot weather.

Treatment.—*Veratrum and Phos.-acid*, given alternately, at intervals, as frequently as the discharges from the bowels occur, will generally be sufficient. If there is nausea or vomiting, or cramping pains in the bowels, give *Ipecac.* in alternation with one or both the former. If thirst and a burning of the stomach or bowels exist, use *Arsenicum*. This last medicine may be given in alternation with either of the others, but is most frequently indicated in connection with *Veratrum*. The intervals between the doses should be regulated by the frequency of the evacuations in all cases, lengthening them as the evacuations become less frequent, until they cease.   In *children*, where the discharges are greenish or slimy, and contain undigested food, give *Chamomilla* and *Ipecac.* alternately, as above directed.. If the discharges are dark, or yellow, with distress in the stomach, give *Podophyllum.* The dose is from 3 to 6 pellets. In all cases of diarrhœa, adults should abstain from all kinds of food until cured, if possible, and eat but little at first, when food is taken. Children should be fed carefully, and but a small quantity at a time, being particular both for adults and children to use as little *liquid* as possible ; drink water in *small* quantities, not very cold.   Avoid exercise, and lie on the back quietly, when that is practicable.   In a large majority of cases, *Veratrum*, if given in the early stages of the disease, will arrest it at once, and in many chronic diarrhœas of weeks or months standing, it is the surest remedy.   In chronic diarrhœa of females, *Podophyllum* should be used in alternation with *Veratrum.* [In many cases of dysenteric-diarrhœa *Baptisia* acts very well, particularly if there is much debility.]

Dysentery.—This disease is caused by inflammation of the mucous membrane of the colon and rectum, generally confined to the lower part of the bowel. It is always painful. There is griping and straining in the lower part of the abdomen, and generally great bearing down when at stool,with a peculiar distress after the evacuation, called tormina. The discharges often commence like a common diarrhœa, with copious liquid evacuations, but there is more or less griping pain, low down, from the beginning. The evacuations sooner or later become lessened, slimy or bloody, or both, the pain increasing accompanied with more or less fever, often quite severe. Sometimes the patient is costive, and has been so for several days, the dysentery coming on without being preceded by looseness. At others, especially in summer, when fevers are prevailing, the dysentery begins with a severe chill, followed by fever and the dysenteric symptoms above described.

TREATMENT.—If it begins with looseness without blood, give *Arsenicum* and *Veratrum* alternately, once an hour, or oftener if the evacuations are more frequent. If the discharges are bloody, use *Mercurius-cor.* in place of the *Arsenicum.* If there is any sickness of the stomach, or the discharges are dark or yellow, use *Podophyllum* with *Mercurius-cor.* If there are colic pains in the bowels, use *Colocynthis* alternately with the others, giving it between them. If the patient was costive previous to the attack, and the dysentery came on without much looseness, *Nux - vomica* should be given alternately with *Mercurius-*

2

cor. If the disease comes on with a chill, or a chill occurs at any time during the attack, followed by fever, *Aconite, Baptisia* and *Podophyllum* should be used in rôtation half an hour apart until a free perspiration is produced, and the pain diminishes; or if bloody stools appear, use *Mercurius-cor.* with the *Aconite* and *Baptisia.* [If the evacuations are principally blood give *Hamamelis viry.*]

A large proportion of the dysenteries of hot weather in miasmatic regions, will be arrested in a few hours by these three or four remedies, especially if the patient keeps still, and generally even if he keeps about his business. In very bad cases, much benefit will be derived from injections of Gum Arabic water, or mucilage of Slippery Elm thrown into the bowel in quantities of a pint or more at a time, as warm as can · possibly be endured. I have often relieved patients immediately with injections of a strong solution of Borax and rice water, as hot as bearable. *Never apply cold water* to *any* inflamed surface, much less a *mucous* surface. All food should be withheld as far as practicable and not starve, until the symptoms abate.

Colic.—The symptoms of this are cramping pains in the abdomen, without fever or looseness of the bowels. The colic sometimes occurs after the cessation of a diarrhœa that had been induced by severe cathartics. The pains are cutting and straining, drawing the bowels into knots, relieved temporarily by pressure.

TREATMENT.—For a male, *Nux-vom.*, and for a female, *Pulsatilla* will generally afford immediate relief. In children, especially where diarrhœa exists, *Chamomilla* should be used. If it is the result of severe cathartics, or if there is a soreness or a bruised feeling, *Colocynth* is the remedy. Hot injections into the. rectum, and large quantities of warm water taken into the stomach, will often *cure colic.*

Bilious Colic.—This disease, in addition to the symptoms of cutting, cramping pains in the bowels, as in common colic, has great distress in the stomach, with nausea and vomiting, the bowels being costive, the feet and hands cold, sometimes cold sweats occur. There is also considerable fever, and frequently headache is present. The substance vomited is at first dark bilious matter, but if the case continues a long time, stercoraceous (fecal) matter will be thrown up.

TREATMENT.—*Colocynth* is the most important remedy, and should be given early and constantly. *Podophyllum* is next in importance, and it should be given in alternation with the former, the dose to be repeated as often as every half hour at first, and as the patient becomes easy, at longer intervals. In this, as in the former case, great benefit will be derived from large injections of quite warm water, and let it be taken into the stomach

freely, as hot as can be safely swallowed. I
have given a gallon of hot water in the course
.of two hours, to a patient suffering under this
disease. the first half pint being rejected, but
the balance remaining, perfect relief having
been experiénced. If fever continues after the
colic and nausea cease, *Baptisia* and *Aconite*
should be given alternately every hour until
the fever subsides. If the patient is, and has
been. for some time, costive, *Nux - vomica*
should be given once in six or eight hours until
the bowels move. Injections may also be
used.

Cholera Morbus.—This disease generally
comes on at night, in hot weather, and is, in
many cases, induced by over-eating while the
patient is suffering from diarrhœa and a deranged state of the liver. It is essentially of
a bilious character. It sets in with great pain
in the bowels, sickness at the stomach, and
vomiting of large quantities of dark greenish
bitter-tasting substance. At first. the vomiting will seem to afford relief, but sooner or
later the stomach and bowels cramp, and the
cramping may extend to other parts of the
body. the feet, hands, calves of the legs, and
the arms, cold sweats come on, and death terminates his sufferings.

TREATMENT.—*Ipecac.* and *Colocynth* are to
be given in alternation, and repeated as often
as every 30 minutes, for the first three or four
doses, then as the patient gets easier, at longer

intervals. A dose every hour will suffice as soon as the symptoms begin to abate. The application of hot cloths or even mustard, over the abdomen, frequently palliates the sufferings, and does not interfere with the action of the medicines. Fever of a low typhoid type sometimes sets in after an attack of cholera morbus, and terminates fatally. This ought never to occur under homœopathic treatment. For such fever give *Baptisia*, a dose every hour until the fever subsides, which will occur generally in six or eight hours; if not, and the patient complains of headache, or is delirious, or dizzy, or feels a fullness in the head, give *Cimicifuga* in alternation with the *Baptisia*. Keep the patient very quiet and free from noise, as far as possible. *Sleep* is a great restorer in any case, but particularly so in this. [*Veratrum-album* is often eminently serviceable.]

FEVERS.—Intermittent Fever, Ague or Chill Fever.—This comes on with pains in the head and back, aching in the joints, yawning, followed by coldness of the hands and feet, blueness of the nails and skin of the hands, general chilliness, sometimes "shaking." This lasts from a few minutes in some cases, to several hours in others. The chill is followed by a fever, which is generally severe and long continued, in proportion to the length and severity of the chill. The fever is followed by free perspiration, when it subsides and leaves the patient in a comfortable condition. This state is called the Apyrexia (*Intermission.*) This continues from a few hours to twenty-four, or longer, when another chill comes on followed by fever and sweats as before. During the chill and fever, the patient often suffers great pain, and is sometimes delirious. Young children frequently have convulsions when the chill sets in. *These* convulsions of children, though alarming, are not often dangerous.

TREATMENT.—As soon as the first symptoms of the chills appear, such as, the headache pain in the back and bones, coldness of the hands, nose and ears, give *Aconite* and *Baptisia* alternately, giving the first three doses every ten minutes, the next three doses every fifteen minutes, and then once in ha'f an hour until the patient begins to sweat freely, when the medicines should be discontinued. If there is nausea or vomiting present, let the patient have lukewarm water freely in large draughts, until he vomits it up several times. As soon as the sweating commences, give *Arsenicum* and *Cimicifuga* alternately every hour during the intermission, except during sleeping time. On return of the chill, should it appear a second time, use the *Aconite* and *Baptisia* as before, and follow them with *Arsenicum* and *Nuxvom.* every two hours. This course of treatment will cure a majority of cases, but some require *Cinchona*. That Cinchona is a specific for intermittent fevers in many of their forms, no one will deny. It is the homœopathic remedy for many cases, and should be prescribed. The injurious effects that are often attributed to Quinine, are, I have no doubt, attributable not to that remedy, but to the *drugs* that are used prior to its use. I have used it in more than two thousand cases, and have never been able to see any evil consequences follow its *proper* use. It should be given *from the beginning of the chill to the end* of the paroxysm, and continued during the whole time of the intermission: *i. e.* until the time arrives for the next chill, *time* being important in the use of this remedy. Use the first decimal trituration, and give grain doses (equal to one-tenth of a grain of the drug,) every half hour till the time the next chill would occur, if it pursued its regular course, allowing the patient six or seven hours time in each twenty-four, for sleep. Though from two to four grains of the pure *Chininum-sulphuricum* is all the patient would get, very few cases that do not yield to a course of the former treatment here recommended, will have the third paroxysm after this *China* treatment is commenced and pursued as here directed. For children the dose may be one-half or one-fourth that of the adults. If a trituration of the medicine cannot be got conveniently, four grains of the *Quinine* may be put into a four ounce vial of water slightly acidulated with Sulphu-

ric-acid, shaken well every time, and a teaspoonful taken at a dose. [A much preferable
mode of administration is to give the sugar-
coated pills of Quinine.] Abstinence from
food as far as practicable, and quiet is of much
importance in this disease, but the patient may
use water freely. In some cases, the chill is
irregular and indistinct, the patient is thirsty
during the chill, and the cold stage is long in
proportion to the length of the fever, the surface pale and more or less bloated. *Arsenicum*
is the remedy, and should be given from the
commencement of the chill, and every hour
until the fever subsides, then every three hours
during the intermission. In chronic cases,
where the patient has been drugged with mer-
curials and cathartics, together with larger
doses of Quinine, and is still suffering under
the disease, *Pulsatilla* and *Cimicifuga* in alternation, will, in nearly every case, effect a
cure.

Bilious Fever.—This may be either intermittent, remitting, or continued and typhoid.
It is distinguished from common intermittent,
by the great derangement of the stomach, as
nausea and vomiting of bilious matter, yellow
coated tongue, bitter taste in the mouth, foul
breath, loss of appetite, high colored urine,
and frequently distress and fullness in the
right side, (though this last is not in every
case present,) the skin and white of the eyes

soon become yellowish, the chills are often imperfect, the fever being disproportionably long.

TREATMENT.—*Podophyllum* and *Merc.* should be given in case of intermittents of this character, during the paroxysm, and in rotation with the other remedies for intermittents, giving a dose every three hours during the intermission. It is well also to continue these remedies night and morning, alternately, for a week or so after the cessation of the chills and fever, or until all bilious appearances cease.

REMITTENT FEVER.—A remitting fever is one that goes nearly off, but not so entirely as an intermittent, returning again by a paroxysm of chill more or less distinct, sometimes hardly perceptible, and an increase of the fever following, from day to day, until arrested.

[*Treatment.*—Gelseminum tincture 5 drops in half a tumblerful of water. give in teaspoonful doses repeated with sufficient frequency to produce moisture of the skin. *This remedy alone will cure all grades of fever except the true typhoid.*]

CONTINUED FEVERS are generally of a bilious character, except in winter, when they are more or less connected with irritation of the lungs, or with rheumatic affections, when they are termed catarrhal or rheumatic fevers. If the bilious symptoms prevail, give *Aconite* and *Baptisia* during the chills and high febrile stage, at intervals of an hour, and during the declining stage of the fever, give *Podophyllum*

and *Mercurius* until a perfect intermission is
produced, when the same treatment should be
adopted as in intermittents.

Catarrhal Fever.—The head being "stuffed up," there is pain in the head, the lungs
oppressed, cough and sneezing, the eyes and
nose suffused with increased secretion of tears
and mucus, pain in the back or loins, almost
constant chilly sensations. use in rotation *Baptisia, Copaiva* and *Phosphorus,* giving a dose
every hour until the fever begins to abate and
perspiration comes on, then leave off the *Baptisia,* and give in its stead *Cimicifuga,* lengthening the interval between the remedies to two
hours or longer. For the *chronic cough* that
sometimes follows catarrhal fever, *Copaiva
Macrotin* and *Phosphorus* should be used
morning, noon and night, in the order here
named.

Rheumatic Fever. — (*Rheumatism.*) the
patient complains of soreness of the muscles,
of the chest, back and limbs, with or without
lameness o the joints, *Aconite, Cimicifuga*
and *Nux-vomica,* are the remedies for a male
patient, and the two former, with *Pulsatilla,*
for a female, (or for a male, of light hair, delicate skin, feminine voice and mild temper,) to
be used in rotation one hour apart. These
remedies are to be taken in a severe acute case,
every half hour until the symptoms begin to
abate ; then every hour or two hours as the
case progresses. [In many cases Caulophyllum

in doses of 3 drops of the first dilution every two hours, will produce a curative effect more rapidly than any other medicine.] *Baths* properly administered, are of great importance in all forms of fever. The surface of the patient should be washed and thoroughly *rubbed* in water quite warm, into which a sufficiency of the lye of wood ashes has been put to make it feel quite slippery. This should be done twice daily in all fevers.

Rheumatism.—In addition to the medicines directed under the head of *Rheumatic Fever*, the most decided benefit can be derived from *Alcoholic Vapor Baths*, which, while they do not in the least interfere with the action of the medicines, tend greatly to mitigate the pains, and produce an equal state of the circulation by stimulating the surface ; abridging in many cases, the disease one-half the time it would run under the long interval treatment alone. This is to be applied by filling a tea cup with alcohol, placed in a saucer of water to insure against danger from an overflow while burning. Place both under a solid wood bottom chair, elevated about the thickness of a brick under each post, strip the patient naked, and after giving him the alkaline bath, and rubbing his surface dry, place him upon the chair, enveloping him completely, except his head, with a woollen sheet or blanket, (as there is no danger of the wool taking fire,) letting the blanket enclose also the chair and

come down to the floor. Then set fire to the
alcohol, and if the heat is too great, raise the
edge of the blanket and let it become reduced.
Continue this until he sweats freely, or becomes too much fatigued to sit longer. Let
the patient often drink freely of cold water,
during the process. Remove him from the
chair to his bed and cover him warmly. It
is well to place the feet in hot water during
this process. This is a delightful operation
for a rheumatic patient, and no one will object
to a repetition of it. Whatever physicians
may think or say of this operation, I *know* it
is a most potent agent for the *cure of inflam-*
matory rheumatism. and is a valuable agent in
the chronic form of this disease.

Typhoid Fever.—This is a dangerous, and
with ordinary allopathic treatment, a very fatal
disease. It generally comes on insidiously,
the patient feeling a dull headache. more or
less pain in his joints, back and shoulders,
with loss of appetite, restless and disturbed
sleep, slight chilly sensations, with a little
fever, dry skin, and a general languid feeling.
These symptoms continue from four or five
days in some cases, to two or three weeks in
in others, gradually getting worse until the
patient is prostrated. or if he takes no drugs,
and keeps still, avoiding food as far as practicable, he may escape prostration, and after
lingering for eight or ten days. and sometimes
longer, just on the point of prostration, he be-

gins slowly to get better, and recovers about
as slowly and imperceptibly as he grew sick.
This is in accordance with observation of cases
under my own eye, and I have no doubt those
cases of spontaneous recovery, had they taken
a single dose of active cathartic medicine or
any of the active drugs, they would have been
immediately laid upon a bed of sickness from
which a recovery would have been extremely
doubtful. I believe that two-thirds of the
deaths from typhoid fever are the direct results of medication, and that those who recover, do so in spite of the active drugs when
such are used. Some cases, however, will not
thus spontaneously recover, and require proper
treatment; and it is safest to treat all cases,
at as early a day as possible. Some cases
come on more rapidly and run into the prostrating or critical stage, in a very few days.
Delirium is a symptom that comes on early in
these cases. When the disease is fully estab-
lished, and even sometimes in the early stage,
diarrhœa sets in and runs the patient down
rapidly.

TREATMENT.—In the early stage, that which
might be called premonitory, while the patient
is yet able to be about his business, but is
complaining of the symptoms above named,
he should, as far as possible, abstain from exercise and food, and take of *Baptisia* and *Phosphorus* alternately, a dose once in three hours.
These will almost invariably produce amend-

ment in a few days, and as soon as he improves
any, leave off the medicines. Should there be
diarrhœa present, use *Phos.-acid* instead of
Phosphorus. If the patient is delirious or has
fullness and redness of the face, the eyes red,
and headache, give *Belladonna* in rotation
with the other two. For the foul breath that
comes on, use *Mercurius-cor.*, especially if the
diarrhœa assumes a reddish tinge, like beef
brine. Should the fever at any time rise high,
the pulse being full and hard. give *Aconite*, but
it rarely happens that Aconite is useful in the
later stage. If the patient complains of pains
in the back, and fullness in the head, give *Cimicifuga.* This is particularly useful for persons who have rheumatic pains in the limbs or
back, during the fever. If the evacuations from
the bowels are dark, or yellow and consistent,
or there is bilious vomiting, *Podophyllum* is
the remedy. From some cause or other. to me
wholly unaccountable, the writers generally
have laid down Rhus and Bryonia as the remedies in typhoid fever. I must confess I
have no confidence in them for this fever as it
prevails, and has for several years past. in this
country. I am confident, from thorough trial,
we have much more reliable remedies in *Podophyllum, Baptisia* and *Cimicifuga.* In the
early stage, or at any time to arrest febrile and
inflammatory symptoms, the *Baptisia* is much
more potent than Aconite, its symptoms cor-

responding peculiarly with typhoid fever. It is important to bathe in this disease.

Scarlet Fever.—*Scarlatina.*—This fever assumes two principal forms : Simple or mild, and malignant. In the *Simple form*, there is great heat of the surface, extremely quick and frequent pulse, headache, and some sense of pain and soreness in the throat. After a day or two, there appear upon the surface, bright scarlet patches, in some cases extending over the whole limbs, the skin smooth and shining, and somewhat bloated or swollen ; upon pressure with the finger, a white spot is seen, which soon disappears on removal of the pressure. As the disease subsides, the cuticle comes off (*desquamates*) in patches. In the simple form of this disease, the throat, though often more or less sore, does not ulcerate. In some cases, notwithstanding the fever is high, the pulse frequent, and the throat sore, there may be no external redness, but the mouth and tongue will have a scarlet hue, indicating the existence of disease more dangerous than when it appears externally. *In the malignant form*, the same symptoms are present, the patient suffers more pain in the head; the back and throat, root of the tongue, tonsils and soft palate become ulcerated, turn black, and sometimes gangrenous, proving fatal in a few days, or slough out in large portions, the ulcers destroying the parts extensively. The breath becomes foul and fetid, and the effluvia from the ulcerated surface is very sickening to the patient and all around him. This disease rarely attacks adults, but occasionally, and for the last six or eight months, in one region where I am acquainted, where Scarlatina of a malignant type has prevailed among children, adults have been affected with an epidemic soreness of the mouth and throat, strongly resembling the worst form of the *angina* in malignant Scarlatina, together with a low typhoid form of fever.

TREATMENT.—In simple scarlatina, all that is necessary is to keep the patient quiet, in a room of uniform temperature, as far as practicable ; let the drink be cold water only, and give *Aconite,* *Belladonna* and *Pulsatilla* in rotation, a dose every hour until the fever subsides. If any soreness of the throat remains, give a few doses of *Mercurius-sol.* If the fever subsides, and the soreness remain, *Hydrastis* or *Eupatorium-arom.* will soon complete the cure. In the *malignant* form, with ulcerated, dark colored, or red and purulent throat, and typhoid form of fever, give *Aconite* and *Belladonna* in alternation, every hour, and at the same time gargle the throat freely with *Hydrastis.* Some of the tincture may be put in water, about in the proportion of ten drops to a teaspoonful, or a warm infusion of the crude medicine may be used. This can be applied with a camel's hair pencil, or a swab, to the parts affected, once in two hours, and

will soon bring about such a state as will result in speedy recovery. After the active fever has subsided, the *Aconite* and *Bell.* may
be discontinued, and *Eupatorium-arom.* used
instead, once in three hours until convalescence is complete. With these remedies, applied as here recommended, my brother, Dr
G. S. Hill, has treated a large number of malignant sore-throats, ("Black tongue Erysipelas,") and been universally successful, relieving them in a few hours, when the symptoms
were of the most alarming character, and the
disease in some cases, so far advanced that the
patients were considered by their friends and
attendants, "at the point of death." The *Hydrastis* is a most potent remedy in putrid ulcerations of the mucous surfaces, and much
the same may be said of *Eupatorium-aromati-
cum.*

Yellow Fever.—As I have never practiced
farther South than Cincinnati, and have seen
but few cases of this disease, my experience
with it has not been sufficient to be relied
upon as authority. Therefore, I shall give a
brief description of the disease, with the proper and *successful treatment*, furnished me by
A. H. Burrett, M. D., of New-Orleans, who is
not only a physician of more than ordinary
learning and skill in his profession generally,
but is one who has spent his time in New-
Orleans among the sick of Yellow Fever,
through three of the most fatal epidemics that

ever scourged any city. He is a man for the times, a man of resources, who draws useful lessons from experience and observation. Hence he has been able to select such remedies as have enabled him to cope most successfully with the pestilence, saving nearly all his patients, while, under other treatment, a majority have died. I therefore, attach great value to his treatment, and recommend its adoption with the most implicit confidence.

"When this Fever prevails as an epidemic as it usually does, in the southern part of the United States, it is a disease of the most malignant character. The proportion of *fatal* cases under the allopathic course of treatment has been equal to, and, in some places. as in New-Orleans, and some towns in Virginia, has exceeded that of *Asiatic* Cholera. It is almost entirely confined to Southern regions, and only prevails in hot weather, after the continuance of extreme heat for some weeks.

It usually begins with premonitory symptoms somewhat like those of ordinary fever, but with this difference : the patient, instead of losing his appetite, has often a morbidly increased desire for food. He complains of severe pains in the back, and more or less headache. Both the head and backache are of a peculiar character : the pains resembling rheumatic pains, the head feeling full and too large, the eyes early turn red, almost bloodshot and watery, a chill comes on, which may be distinct and quite severe, lasting for an hour or more, or, it may be slight, and hardly perceptible. The chill is followed by high fever, tho pain in the head and back increasing, the eyes becoming more red and suffused, the forehead and face extremely red and hot, and the heat of the whole surface very great, the carotids beat violently, the pulse very frequent, and usually, at first, full and strong, though sometimes it is feeble from the beginning. However the pulse may be in the beginning, it very soon becomes small, but continues to be frequent. The tongue is at first covered with a white paste-like coating, which afterwards gives place to redness of the edges and tip with a dark or yellow streak in the centre. The stomach is very irritable, rejecting every kind of food, and all drinks, except, perhaps, a few drops of ice water. There is a peculiar distressed feeling in the stomach, often a burning sensation, so that, if suffered to do so, he would take large quantities of ice or water. One remarkable feature of the case noticed in the epidemic, as it existed in New-Orleans the past season, was, that the patients had a great desire for food, notwithstanding the nausea and distress at the stomach.

Sooner or later, varying from a few hours to several days, in the ordinary course of the disease, the fever subsides. From this time the patient may recover without any further symptoms, but this is, by no means, the usual result. If the subsidence of the fever is accompanied by natural pulse, a free, ·but not profuse or prostrating perspiration, a genial warmth of the surface,· natural appearance of the countenance, eyes, and tongue, with little or no soreness on pressure over the stomach, we may safely look for a speedy recovery. But if, on the contrary, the eyes, face, and tongue, become yellow, or orange-colored, the epigastrium is tender to pressure, the urine has a yellow tinge, the pulse becomes unnaturally slow, with the least degree of mental stupor, we have reason to know, full well, that the lull of the fever is only the calm preceding a more destructive storm. The fever has subsided, only because exhausted nature could re-act no longer. It may be in a few hours, or not until twelve or twenty-four have elapsed, the pulse becomes quickened, even to the frequency of 120 to 140 in a minute, but very feeble, the extremities of the fingers and toes turn purple or dark, the tongue becomes brown and dry, or is clean, red, and cracked, sordes may be on the teeth, the stomach becomes more irritable, nausea and vomiting are extreme, the substances vomited being, at first, reddish, afterwards watery, containing flocculæ, like soot, or coffee grounds; the breath becomes foul, and the whole surface emits a sickening odor. The pulse becomes very small, though the carotid and temporal arteries beat violently. The urine fails to be secreted, and later, blood is discharged from the mucous surfaces, involuntary discharges from the bowels, clammy sweats; and death follows.

The disease runs its course in from three to seven days, sometimes proves fatal in less than a day, and at others, assumes a typhoid form, and runs for weeks. Occasionally it sets in without any of the premonitory symptoms, the chill being first, the fever following, succeeded immediately by the black vomit, going through all the stages in a single day, or two days.

Again, it sometimes begins with the black vomit, the patient being immediately prostrated. In all cases, however it may begin, the peculiar head-ache and back-ache as described in the beginning, as well as the extreme heat of the head and face, redness of the eyes, the gnawing sensation at the stomach, and peculiar nausea are present. These seem to be characteristic symptoms that mark the Yellow Fever, and those which should guide in the search for the proper remedies.

TREATMENT. — The remedies that proved successful in arresting the disease during the early or forming stage, before the chill or fever had set in, while the symptoms were pain, fullness, and throbbing of the head, with more or less dizziness, rheumatic pains in the back, and redness of the eyes, were *Aconite* and *Bell.*, at low attenuations, once in two to four hours, according to the violence of the symptoms. For the fullness of the head, pressing outwards, as though it would split, with pains of a rheumatic character (*Cimicifuga* 1st, given in one grain doses, every hour or two hours, proved specific. These three remedies *Aconite*, *Belladonna* and *Cimicifuga*, would in nearly all cases, arrest the disease in the forming stage, so that no chill or fever would occur, or, if fever did come on after this treatment, it was mild.

When the fever sets in, and the pain in the head and back increases, the eyes, forehead and face are extremely red, or purple and hot, the pulse frequent and full, the tongue coated white, *Aconite*, *Belladonna* and *Cimicifuga* are still to be relied upon, but they should be given every half hour, in rotation, at low attenuations. If the tongue is red, in the early stage, use *Bryonia* in place of the *Belladonna*. In a later stage, when sickness or distress at the stomach had become prominent, with the quick pulse, and hot skin, *Ipecac.* and *Aconite*, both at the 1st attenuation, a dose given every half hour alternately, generally arrested the symptoms, and brought on perspiration of a healthful character, followed by subsidence of the fever and convalescence. Sponge baths, with half an ounce of *Tr. Ipecac.* in two quarts of tepid water, applied to the whole surface freely, under the bed clothes, so as not to expose him to the air, contributed much towards bringing on perspiration and subduing the fever, as well as allaying the nausea.

When called to patients in the stage of *Black Vomit*, whether that came on as an early symptom, or at a later stage, *Nit.-acid*, *Vera'rum-viride* and *Baptisia*, all at the first dilution, were administered every hour in rotation, with great success, the symptoms yielding in a few hours. For the great oppression, as of a load in the stomach, without vomiting, *Nux* was found sufficient. In the later stage, when there seemed to be no secretion of urine, *Cannabis-s.* and *Apis-mel.*, gave relief.

The remedies most successful for the cases that assumed a typhoid character, with dry, cracked tongue, sordes on the teeth, and low sluggish pulse, were *Baptisia* and *Bryonia*, given every two hours, alternately. *Nitric-acid* given internally and injected into the rectum, when bloody discharges appear, is generally quite successful.

Good nursing is of the utmost importance, and the patient should be visited frequently by his physician, as great changes may occur in a short time. Three times a day is none too often to see the patient. As soon as the fever comes on, the patient should be stripped of his clothes, and dressed in such garments as he is to wear in bed through the attack.

He should be put to bed and lightly covered, but have sufficient to protect him from any sudden changes in the atmosphere, and the room should be well ventilated all the time. The baths should always be applied under the bed clothes.

The diet should. be very spare and light, after the fever subsides, and while the fever exists no food should be taken. Thin gruel, in teaspoonful doses, once in half an hour, is best. After a day or two, the juice of beef steak may be given in small quantities but give none of the meat. No "hearty food" should be allowed for eight or ten days after recovery. A relapse is most surely fatal.

As *prophylactics* (*preventives*) of the fever, *Cimicifuga, Bell.* and *Aconite* should be taken, a dose every eight to twelve hours, by every one that is exposed. These will, no doubt, often · prevent an attack, and if they do not, they will so modify it, that it will be very mild, of short duration, and very easily arrested.

Pregnant females, and young children were sure to die if attacked, when treated by the allopathic medication ; but, by the use of these remedies as *preventives*, their attacks were rendered so mild as to be amenable to remedies, and all recovered.

Hepatitis (*Inflammation of the liver.*)— SYMPTOMS.—Pain in the right shoulder and loin, weight and tension in the region of the liver, tenderness on pressure, nausea, vomiting, coated tongue. loss of appetite, deranged bowels, thirst, fever, dry cough. depression of spirits, bowels constipated or relaxed, urine scanty and bilious. When the upper surface of the liver is affected the cough is more troublesome and inspiration painful. When the lower surface is inflamed there will be a prominence of gastric disturbance. *Chronic hepatitis* comes on insidiously, with symptoms of dyspepsia. obtuse pain in the region of the liver, sallow skin, despondency, clay-colored stools, sometimes diarrhœa. sourness of stomach, colic pains, disposition to nausea. white tongue, wasting of flesh, dryness of skin and slight disposition to fever.

TREATMENT.—Give *Aconite* and *Baptisia* in alternation every hour during the high febrile stage for twelve to twenty-four hours or until a free perspiration is sooner produced. If nausea or vomiting is present, encourage these symptoms by copious draughts of lukewarm water, until the stomach is thoroughly evacuated, say, until the patient has three or four spells of free vomiting. Then use the *Aconite* and *Baptisia,* giving one-drop doses of the latter in pure tincture, and the former at the second dilution, letting the patient drink freely of *hot* water, as hot as can be safely swallowed, into which a little sugar and milk is put, to make it more palatable. After twelve hours treatment as above, or in two hours after free perspiration is produced and has been kept up, give *Podophyllin, Mer.-sol.,* and *Leptandrin* in rotation, second decimal trit., of each in one-grain doses (for an adult) a dose every two hours, for the first day. Except during sleep, which should never be interrupted for the propose of giving medicine. gradually lengthening the interval as the patient improves. until convalescence · is complete.

For **Chronic Hepatitis** and symptoms of *Jaundice* , that sometimes follow an acute inflammation of the liver, I have been abundantly successful with. *Podophyllin, Leptandrin, Mer.-sol.* and China, at the second attenuation, given one dose of each daily, about four hours apart, taking care to have them come at least one hour either before or after meals. It is better to be given before meals, if practicable, except one dose at bed-time. Let the *Podophyllin* be given at bed-time.

For chronic affections of the liver bathing the surface daily in weak lye. warm in a warm room, and rubbing the surface freely, and this followed by sponging the surface with equal parts of alcohol and water, in which *Quinine* at the rate of forty grains to the pint has been dissolved, will greatly aid in the cure. This is especially useful in regions where ague or intermittent and typhoid fevers are prevalent.—It acts like a charm—especially with females who are suffering from general debility, as well as torpidity of the liver.

Pleurisy - Pleuritis.—This is an inflammation of the pleura of one or both lungs, generally confined to one side. It is known by sharp pain in the side of the chest increased by taking a long breath, or coughing, or by pressing between the ribs. The cough is dry and painful, the patient makes an effort to suppress it, from the pain it gives him ; the fever is of a high grade, the pulse full, hard and frequent, with more or less pain in the head.

TREATMENT.—*Aconite* is a sovereign remedy. It should be given at intervals proportionate to the severity of the disease, once in half an hour, for about three doses, then every hour until the patient is easy and perspires freely. This is the course I have generally pursued, and scarce ever failed of relieving in a few hours. [*Bryonia* is advantageously alternated with *Aconite*.] Other means may often be used with advantage at the same time, and not interfere with the action of the medicine. Put the feet and *hands* into water as hot as it can be endured, and apply to the affected side very hot cloths, hot bags of salt, or mustard. There is no harm in this, and it relieves the pain. Let the patient drink freely of *hot* water, into which you may put milk and sugar to render it palatable. If the case seems to linger, and perspiration is tardy in appearing, give, in alternation with Aconite, *Eupatorium-arom.* This will soon relieve.

[*Hæmorrhage* from the lungs (*Hæmoptisis*) yields promptly to *Arnica* and *Hamamelis-virgin*. Five drops of tincture to one-quarter pint of water, a teaspoonfull every fifteen minutes. If much fever give Aconite same manner. If debility follows give Phosphoric-acid.]

Inflammation of the Lungs—Pneumonia.

This disease is often connected with Pleurisy, and consists of inflammation of the substance of the lungs. As in the former case, it may attack only one, but may exist in both sides at the same time. If the pleura is also affected, there will be all the symptoms of pleurisy, together with those peculiar to inflammation of the lungs proper. They are, pain in the lungs, oppressed breathing, cough, causing great distress on account of the soreness of the affected parts: at first, expectoration from the lungs is nearly wanting, the cough being dry, but after a time, there is a rattling sound on coughing, and more or less mucous substance is with difficulty raised. This is, at first, white or brownish, but soon becomes reddish and frothy, tinged with blood. *The patient lies on the affected side, and cannot rest on the sound side.* The pulse is full, hard and frequent, the fever high, pain in the head, and sometimes delirium. If the disease is not arrested, the patient generally dies from suffocation, by *the lungs* filling up, hepatized, or abscess and ulceration come on, and then what is called "quick consumption" carries him off.

TREATMENT.—In the early stage, *Aconite* and *Phosphorus* should be used at intervals of from half an hour to one hour, in alternation, until the fever abates, and the oppression in the chest is relieved. If, however, there is bloody expectoration, *Bryonia* may be used in place of *Phosphorus*, though I prefer to use it in rotation with the two others. These will soon, in all ordinary cases, subdue the most distressing symptoms, and effect a perfect cure in a day or two. *Belladonna* should be used, when there is much delirium, or great pain in the head. Occasionally, the cough from the beginning, is apparently loose; there being a rattling sound, but the expectoration is difficult, the fever high, with some chilly sensations, or at least, coldness of the knees, feet and hands, a white or brownish fur upon the tongue, and pain in the bowels For such symptoms, especially with the pain in the bowels, as though a diarrhoea would come on, give *Tartar-emet.* It is often one of the best remedies in this disease, affording relief when others have failed. After subduing the high febrile symptoms, if there remains cough, indicating much irritation, or inflammation of the lungs, *Cimicifuga* should be used in place of Aconite, with *Phosphorus* and *Copaiva*, the three in rotation, two hours between doses.

Acute Bronchitis.—*Inflammation of the Bronchial Tubes.*—This is attended with distressing cough, profuse expectoration, oppressed breathing, pain in the forehead. and general catarrhal symptoms. *Baptisia. Copaiva* and *Eupatorium-arom.* given every hour, in rotation, will, in general, relieve from the acute affection in a short time.

Chronic Bronchitis requires the use of *Copaiva, Cimicifuga* and *Arum-triphyllum,* to be taken morning, noon, and night in the order named ; or, if the cough be severe, they should be used every three hours. These will be sufficient to effect a cure.

Coughs generally, unless they arise from from consumption, yield readily to the alternate use of *Copaiva, Phosphorus* and *Cimicifuga,* a dose given once in from three to six hours. If, however, there is soreness of the throat, redness and soreness of the tonsils, palate, and fauces, or soreness of the larynx, with hoarseness, *Arum-triphyllum* and *Hydrastis-can.* are the surest remedies. They rarely ever fail of effecting a complete cure in a few days. They should be used three or four times a day. They may be used with the other medicines recommended for coughs. [*Sticta-pulmonaria* is a very efficient remedy for catarrhal coughs, particularly if accompanied by rheumatic pains.]

Acute Sore Throat, arising from sudden cold, *Arum-tryphillum,* and *Eupatorium-arom.*

are the remedies to be relied upon. If the tonsils seem to be mainly involved, constituting

Quinsy—Tonsilitis, *Belladonna* and *Aconite* should be given, while there is high fever, then substitute for them *Arum-tri.* and *Phos*, or these may be used in rotation with the former, a dose every hour or oftener.

Inflammation of the Bowels - Enteritis. —This consists in inflammation of the muscular and peritoneal coats of the intestines, sometimes also involving the mucous coat. The pain in the abdomen is constant, intense and burning in its character, felt most at the navel; the abdomen is extremely tender to pressure, and often bloated or tympanitic. Thirst is intense, but cold drinks distress and vomit the patient. The pulse is small, feeble and frequent, and the bowels costive. This is a very dangerous disease. It is sometimes connected with inflammation of the stomach, then called gastro-enteritis. The tongue is then red and pointed, the nausea and vomiting are more violent and constant, the thirst burning and insatiable.

TREATMENT.—The same medicines are applicable to both *Gastritis* and *Enteritis*. *Aconite*, *Arsenicum* and *Baptisia* should be used one following the other every half hour until the symptoms begin to subside, then let the intervals be lengthened. In addition to these remedies, I allow the patient to drink often and freely of hot water, as hot as can be swallowed, and though it is at first almost instantly rejected by the stomach, by repeating it in a few minutes in moderate quantities, it gives relief and will soon so allay the iritation as to remain. In some cases the vomiting is severe, the bowels are loose, and pain burning. For such symptoms *Tart.-emet.* is the proper remedy. Cold drinks should not be taken. Cloths wet in cold water, ice water if it is at hand, and wrung out so as not to drip, should be laid over the whole abdomen and instantly covered with two or three thicknesses of warm dry flannel, and the patient's feet kept warm. This may be considered harsh treatment, but there is no danger in it; on the contrary I have, in the worst and most alarming cases of *gastritis* and *peritonitis*, made such applications. and in less than an hour have seen my patient easy and beginning to perspire freely, all danger having passed. It always affords more or less relief and is never. attended with danger. Covering the wet cloths immediately with plenty of dry ones is very essential. After the acute inflammation has. subsided, it is well to have the bowels moved. but don't give drastic cathartics. *Nux-vomica* given at night and repeated morning and noon, will generally serve to cause an evacuation. Injections may be used.

Croup.—This is a disease of children; comes on in consequence of a sudden cold.

Children suffering from hooping cough are more subject to it. The cough is of a peculiar whistling kind, like the crowing of a young chicken, with rattling in the throat and difficult breathing, fever is present, and often very violent. It is properly an inflammation of the larynx, but the inflammation may also exist in the pharynx. the tonsils may be involved, and it may extend to the trachea. (wind-pipe). A false membrane forms in the larynx if the disease is not arrested, and so obstructs the breathing as to cause death from suffocation.

TREATMENT.—Give at first *Aconite, Phosphoric-acid* and *Spongia.* giving them in the order here named, once in ten minutes in a very violent case. and as the patient improves, at intervals of half an hour, and then an hour. Should the fever subside, and still the tightness in the throat and cough continue to be troublesome, give *Ipecac.* in place of Aconite. And when the cough seems to be deep-seated use *Bryonia* instead of Spongia. The patient should be kept in a warm room. and free from exposure to currents of cold air. The application of a cloth wrung out of cold or ice water to the throat, covered immediately with dry warm flannels so as to exclude the air from the wet cloth, will often exert a decidedly beneficial effect, and there is no danger if managed as here directed. The feet should be kept warm and the head cool, but *don't* put *cold* water on a child's head.

Diphtheria.—This is a disease incident to young persons and children, rarely attacking middle aged or old persons.

SYMPTOMS —Its peculiar characteristic consists of an exudation on the uvula, soft palate or tonsils, of a whitish or grayish color. Yet it is not a local malady but a constitutional affection. It generally commences with a chill, then fever, husky voice, sore throat, frequent pulse and rapid decline of strength. The prostration is a prominent feature of diphtheria In many cases typhoid symptoms arise. The breath is putrescent. This fœtor is almost always present.

TREATMENT.—In the early stage while there is high fever, I give *Aconite* and *Baptisia* in low dilutions, giving equal to one drop of tincture of Baptisia at a dose in alternation, every ten to fifteen minutes, with Aconite 3d. until free perspiration is produced. As a general rule it will take between two and three hours to accomplish this.

If the patient has any nausea, as is often the case, and in fact whether there be nausea or not. it is best to precede the medicines by copious draughts of lukewarm water repeated every five to ten minutes, until he vomits freely two or three times ; then give the medicines as above. to promote perspiration and reduce the fever. I have found nearly all cases to be of an intermittent or remittent type —as to the fever, and that at each subsequent return of the fever all the diphtheriac symptoms were more severe or aggravated. In view of this character of the disease I have given Quinine freely in doses of from a half to three grains, repeated every two hours until the patient has taken from five to fifteen grains, according to the age or susceptibility, or until the ringing in the ears and the fullness and dizziness peculiar to Quinine is produced, showing the system to be under the influence of the medicine. I give the *Bin-iodide* of *Mercury* 2d, in doses of one-fourth to two grains, once in from two to three hours, varying according to the age of the patient and stage of the disease, giving it more frequently in the early stage. It is well to swab out the throat thoroughly, as often as every four to six hours with strong salt and vinegar, used as hot as practicable. The throat should be wrapped in warm flannels and the practice, common among the people, of putting a slice of fat salt pork on the throat bound on by flannel cloths, does no harm, and I am of the opinion that it is positively beneficial. I never forbid, but encourage such applications. If the case is far advanced when first seen I use *Mer.-cor.* 2d, every two hours for 6 to 8 hours before giving the *Bin.*, or else give them in alternation every hour for that time ; then discontinue the *Mer.-cor.* and continue the Iodide.

During the winter of 1862–3 while in the city of Lansing, the author treated over one hundred cases of this disease. The first case he saw was in a very advanced stage and died. All the others recovered. In all these cases, without an exception, the Quinine was freely used, just as soon as the fever began to decline and perspiration had been produced. After the patient was brought fully under the constitutional influence of this medicine the case was easily managed. Nothing more was necessary in any of the cases than a free use of the Bin.-iodide of Mer. 2d, with occasional swabbings and gargles with salt and vinegar, to complete the cure. In a few cases when there were head symptoms indicating *Belladonna* or *Opium*—one of these was given until the symptoms subsided. One remarkable feature of this disease, or its effects, deserves notice. That is the extraordinary debility that continues for weeks, and in some cases, for months after the symptoms all disappear. I found that a few doses of Quinine acted like magic in restoring the strength and removing the peculiar languor complained of.

Asthma.—If an attack comes on from sudden cold, take *Aconite* and *Ipecac.* every hour for a day, and if any symptoms remain, in place of the Aconite use *Copaiva, Arsenicum* and *Phos.-acid* with *Ipecac.*, giving them in rotation, a dose every hour.

In *Chronic Asthma,* where the patient is liable to an attack at any time, great benefit will be derived from taking these four in rotation about two hours apart for a day or two, at any time when symptoms of an attack/begin to appear.

I have recently succeeded in alleviating several bad cases, at once, by these four remedies in succession as here recommended, on whom (some of them) I had at various times tried all of them, as well as other medicines, singly at longer intervals, as directed in the books, without any decided benefit. After trying these in succession, as here directed, I found no trouble in arresting the paroxysm in a few hours, and I am strong in the faith that with some, at least, I have effected *cures*. It is worth much to *arrest* the *paroxysm* if no more.

Pertussis.—Hooping Cough.—This disease may not be entirely arrested in its course, and not generally much abridged in its duration, still the use of appropriate medicines will greatly modify it, and render it a comparative trifling affection.

In treatment. give at the commencement of the attack *Bell.* and *Phos.-acid* alternately every twelve hours for a week. then once in six hours, and if the child should take cold so as to bring on fever, give one every hour. Continue these, as above directed, for the first two or three weeks, then, in their stead, after the cough becomes loose, and the patient vomits

easily, give *Copaiva* and *Ipecac.* in the same
manner as directed, for the two former remedies.

Dyspepsia.—This term is applied so loose-
ly and so indiscriminately to all chronic de
rangements of the stomach, that it is difficult
to define it.   I shall therefore point out some
of the more common ailments of the stomach
and their proper remedies.

For sour eructations with hot, burning,
scalding fluid rising up in the throat, with or
without food, give *Phos.-acid* and *Pulsatilla*
in alternation every half hour, until the stomach is easy. For a feeling of weight and pain
in the stomach, with dull pain in the head,
with or without dizziness, give *Nux.-vom.*
every hour until it relieves.   If there is a
burning feeling in the stomach as well as the
heavy load, *without* eructations and rising of
fluid, *Arsenicum* should be alternated with
the *Nux.-vom.*, at intervals of two hours.
There are persons who, from imprudence in
eating or drinking or both, or which is more frequent. from *harsh drug-medication*, have so
enfeebled their stomachs, that, though by care
in selecting their food, and prudence in taking
it, they may suffer but little, are, nevertheless,
when from home or on special occasions, liable to over-eat or take the wrong kind of
food, from which unfortunate circumstance
they are made to suffer the most tormenting
and intolerable distress in the stomach and

bowels, which may last, more or less severe,
for several days. Soon after the unfortunate meal, perhaps the next morning, or, it
may be, in a few hours, the stomach begins to
bloat, by accumula ing gas within, which is
belched up every few minutes in large quantities; the stomach and bowels are racked
with the most torturing pains: cold sweat
stands on the brow, and he is the very pic-
ture of misery. Thus he may roll and tumble
all night, and remain in misery the next day
and several days longer, before the food will
digest. It often passes from the stomach without digestion, and on its way through the
bowels inflicts constant pain. If he does not
take some emetic substance, he is not apt to
vomit, his stomach cramping so as to prevent
it. I have here described one of the bad cases,
but bad as it is they are by no means *very*
rare. There are such cases in abundance, of
all grades from the one here described down
to a slight derangement. They all require a
similar course of *treatment*. It is useful for
such patients to take at once large quantities
of lukewarm water, and repeat the draught
every ten to fifteen minutes, until free and
thorough vomiting is induced, so as to throw
off all the food from the stomach. But even
this does not often cure these bad cases. If
it did, it is not always convenient to do it.
The medicine that is quite certain to afford
relief at once is *Podophyllum*. Let it be given,

and the dose repeated in an hour. A third dose is rarely necessary. After relief from this attack, the medicine should be taken night and morning for a month or more until the stomach is restored. In the meantime care should be taken not to overload the stomach. [*Pulsatilla* is very efficacious in these cases.]

Constipation.—The medicine for this affection is *Nux-vom.*, to be taken at night on retiring. If there is fullness and pain in the head from costiveness, *Bell.* should be used in the morning. and at noon. [*Collinsonia* affords permanent relief.] Let the patient contract a habit of drinking *cold water* freely on rising in the morning, at least half an hour before eating. The patient *should not take physic* For constipation of children, *Nux* and *Bryonia* are to be given, Nux at night and Bryonia in the morning. *Opium* is useful.

Much needless alarm is often felt by persons on account of a costive state of the bowels. If no pain is felt from it, there is no cause for alarm.

"Heartburn."—This peculiar burning and distressed feeling at the stomach depends on imperfect digestion, but is *not* ordinarily, as is generally supposed, connected with a sour or acid state of the fluids in the stomach. The condition of the fluids is alkaline, in most cases, though it is sometimes acid. If it depends upon biliary derangement, *Nux-vomica* and *Podophyllum* are the remedies for a male; *Pulsatilla* and *Podophyllum* for a female.

Erysipelas.—This is a disease of the skin, producing redness, burning and itching pains, appearing in patches, in adults, most apt to appear about the head and face, but in children, upon the limbs, or in very young children, beginning at the umbilicus. It sometimes begins at one point, and continues to spread for a time, then suddenly disappears, and reappears at some other point. *Simple Erysipelas* only affects the surface, with redness and smarting. *Vesicular,* produces vesicular eruption, or blisters filled with a limpid fluid, somewhat like the blisters from a burn. *Phlegmonous Erysipelas* affects the whole thickness of the skin and cellular tissues beneath it, producing swelling, and not unfrequently, resulting in suppuration, ulceration or gangrene and sloughing of the parts. It is a dangerous disease, especially when on the head

TREATMENT.—For the simple kind, *Bell.* is all that will be needed, unless there should be considerable fever, when *Aconite* should be alternated with the *Bell.* For the *vesicular* kind, where there are blisters. *Rhus-tox.* should be used with *Bell.* For the *Phlegmonous,* with deep-seated swellings, *Apis-mel.* is the most important remedy. I prefer to use three of these remedies, giving them in rotation, beginning with the *Bell.,* followed with *Rhus,* and then by *Apis-mel.* giving them one hour apart. • In a mild case, or after the patient begins to recover give them at longer intervals. The *Apis* alone will often be sufficient. During the whole time, the affected parts should be kept covered with dry, superfine flour, some say buckwheat flour acts most favorably. The diet should be very spare. Eat as little as possible, until the disease begins to subside. A very important part of the treatment of this affection is to keep the patient in a room that is comfortably warm, say at a temperature of from 65 to 75°, and keep the temperature *uniformly the same*, as nearly as possible, night and day. Do not, by any means, expose him suddenly to cold air, or a cold breeze, as on going into a cold room, going out into cold air, or undressing or dressing in a cold room. Uniformly warm temperature is of great importance.

Burns and Scalds.—No matter what the nature and extent of the burn may be, the very best of all medicines of which I have any knowledge, is *Soap*. If the parts affected, are immediately immersed or enveloped in Soft Soap, the pain will be greatly lessened, and the inflammation that would otherwise follow, will be essentially modified, if not entirely prevented. It acts like magic; no one who has never tried it can have any idea of its potency for the relief of pain, together with the prevention of bad consequences following severe burning. Under the influence of the *Soap* applications, burns and scalds will often be rendered comparatively insignificant injuries. Instead of endangering the life of the sufferer from the excessive pain, or the ulceration, or gangrene and sloughing that would follow if the pain in the first instance does not destroy life, the pain ceases, or becomes bearable in a short time, and either little or no suppuration or sloughing takes place, or the sore assumes the appearance of healthy suppuration, and heals kindly — avoiding those unsightly deformities that so commonly follow severe burning. If practicable, the soap, as before suggested, should be applied immediately after the burn, the sooner the better. The part may be put into soft soap, or cloths saturated with it can be wrapped around or covered over the affected surface, to any desirable extent. The parts should not be exposed to the air for a single moment, when possible to prevent it. During the first two or three days, dressings need not be removed, unless they cause irritation after the first severe pain has subsided. They should be kept all of the time moist, and as far as practicable, in a condition to be impervious to the air.

When it is necessary to remove them, let the affected surface be immersed in strong soap suds, at a temperature of about 75 or 80°, and the dressing removed while it is

under water, and others applied while in
the same situation.   In ordinary cases. however, even of extensive burns. after the fever
consequent upon it has subsided. and the
part is tolerably free from pain and smarting,
the dressings may be removed in the air,
but others should be in readiness and applied as speedily as possib'e.   The soap
dressings are to be continued from the beginning until the inflammation has subsided
and the sore has lost all symptoms that distinguish it from an ordinary healthy suppu-
rating sore.

After the first few days, or in case of a slight
burn at the beginning, an excellent mode of
applying the soap, is to make a strong thick
"*Lather*" with soft water and good soap. such
as Castile, or any other good hard soap, as
a barber would for shaving, and apply that
to the affected part with a soft shaving
brush; apply it as carefully as possible, so
as to cover every part of the surface, and go
over it several times, letting the former coat
dry a little before applying another, forming
a thick crust impervious to the air.   In
small burns, and even in pretty extensive and
severe ones, this is the best mode of application, and the only one necessary.

In many cases of very severe and dangerous burns, under the influence of this application, the inflammation subsides, and after
a week or more, the crust of lather comes

off, exposing the surface smooth and well. Although it is important to apply the *soap* early, and the case does much better if that has been done, still I have found it the best remedy even as late as the second or third day. In such a case, the *lather* application is the best.

For the fever and general nervous disturbance, *Aconite* and *Bell.* should be given alternately, as often as every half hour, and the *Aconite* should be given in appreciable doses; it acts powerfully as an anodyne. The soap treatment, or at least, the mode of applying it was first suggested to me by Dr. J. TIFFT, some six or seven years ago, since which time I have had opportunities of testing its virtues in all forms of burns and scalds, some of which were of the severest and most dangerous character, and I am quite sure in several cases, no other remedy or process known to the medical profession, could have relieved and restored as this did.

The application of finely pulverized common salt, triturated with an equal part of superfine flour, acts very beneficially on burns. ' It seems to have the specific effect to " extract the heat," literally putting out the fire. It is particularly useful for deep burns where the surface is abraded. Some may suppose this would be severe and cause too much pain when applied to a raw surface, but so far from that being the case, it is a most soothing application. It often so changes the condition of even the severest burns, in a short time, as to render them of no more importance and no more dangerous than ordinary abrasions to the same extent, by causes unconnected with heat. *Urtica urens* is directed for burns, and is useful, but the *Urtica-livica* is better. [The *Urtica cerate* is an admirable mode of application]

Chilblains—That follow freez ng or chilling the feet, causing most distressing uneasiness and itching of the feet and toes, take these remedies, *Rhus* and *Apis*, the former at night and the latter in the morning. In bad cases, they should be used once in six hours. Applications of *Arnica-cerate* to the affected parts at night, warming them before a fire, will serve greatly to palliate the sufferings, and frequently effect a perfect cure. The *Urtica-divica* will relieve recent cases, immediately, and is one of the best remedies for the chronic affection. It should be taken at the second dilution. and the cerate applied to the affected part every night.

Hoarseness.—This arises generally, from inflammation of the mucous membrane of the *larynx*, in ordinary cases but slight. It is a frequent accompaniment of Bronchitis. The remedies most useful, and those which will, in almost all ordinary cases, remove this affection at once, are *Arum-tri.* and *Copaiva,* to be taken a dose every three hours in alternation. If there is present a dry hacking cough, it will be well to take *Bell.* in the interval between the other medicines, for a day, or until the cough is relieved, or changed to a meist condition. [*Aphonia* (loss of voice) is relieved by *Phosphorus*].

Inflammation of the Brain —*Brain Fever.* —Though this affection is not strictly what is called "brain fever," it is attended with more or less general fever, while in what is called "brain fever," there is great irritation of the brain, requiring in many respects similar treatment. As the treatment proper for inflammation of the brain, with some slight modifications in relation to the existing fever, will be applicable to both, I shall treat of them under one head. Some of the principal symptoms are delirium and drowsiness, fullness of the blood-vessels of the head, beating of the temporal arteries, redness and fullness of the face, the pupils dilated, (though in the very early stage they may be contracted.) If the membranes of the brain be the seat of the disease, the pain is more intense, and frequently the limbs are in a palsied state. The patient sometimes vomits immoderately, and the pulse is slow and irregular, but full. The breathing becomes stertorous. The fever is very considerable, and the head hot.

TREATMENT.—*Aconite, Belladonna* and *Bryonia* should be given in rotation, one dose every hour in a violent case, lengthening the intervals as the symptoms abate. Applying *hot cloths* to the head, removing them occasionally to let the water evaporate, will greatly palliate and will not in the least interrupt the action of the medicines. Never apply cold to the head of any person, when hot or inflamed, much less to that of a child. Children are often killed by the application of ice to the head, producing congestion and paralysis of the brain. Hot applications are homœopathic to the state then existing, and always beneficial. The feet may also be placed in hot water, but children should never be put into a hot or warm bath when sick, so as to cover more than the lower extremities.

Convulsions of Children—Fits—These occur, either from the irritation of worms, or as precursors of ague, or they may arise from diarrhœal irritation, affecting the brain. They sometimes occur in hooping cough. If convulsions occur from worms, the child appearing to be choked, give at once some salt and water, and as soon as the first paroxysm is over, give a dose of *Bell.*, and after an hour a dose of *Santonine*. If they come on at the commencement of an ague chill, give *Aconite* and *Bell.* every half hour for three or four doses alternately, then leave off *Bell.* and give *Baptisia*. If diarrhœa is the cause, give *Bell.* and *Chamomilla*. If from hooping cough, *Bell.* alone should be used.

Measles.—This is a contagious disease, and always begins with symptoms like a cold, with high fever, and a severe dry cough. thirst and restlessness. *Pulsatilla* is the proper medicine to palliate and regulate the symptoms. If the fever is high, *Aconite* should be used every two hours alternately with *Puls.* Should the eruption subside suddenly, give *Bryonia* with *Pulsatilla* until it reappears. Let the child drink freely of cold water, and avoid stimulants of every kind. If the eruption is tardy in its appearance, a hot bath may be administered, being careful to have the room quite warm, and to rub the patient dry, very suddenly after the bath. Frictions by the healthy hand over the surface, will do much towards bringing out measles. After the eruption is out, quiet, freedom from sudden exposure to cold; cold water and light diet is all that is necessary. In some of the most obstinate cases, where the eruptions failed to appear in the proper time, as well as where they had receded too soon, I have been able to bring them out in a short time with Gelseminum, tincture, 5 drops in a half a tumblerful of water, a teaspoonful every half hour. will determine to the skin immediately and bring the eruption out naturally. It is a remedy for measles well worth attention.

Mumps.—This is a contagious disease, consisting in an inflammation of the parotid gland. There is, at first, a sense of stillness and soreness on moving the jaw, soon after the gland begins to swell, and continues to be sore and painful, with more or less headache, and general fever for from six to eight days.   It is not ordinarily a dangerous disease, unless translated to some other part. It may remove from the original seat to the breasts.

TREATMENT.—*Mercurius* should be given three times a day during the attack.   If the brain becomes affected, use *Bell.* and *Apis-mel* in alternation.   Should it recede to the testicles, or to the female breasts, *Apis-mel.* is *the* remedy.   *Mercurius* may be used in connection with the *Apis* as soon as the violent symptoms have subsided, in order to prevent permanent glandular swellings.

Stings of Insects.—The effect produced by the sting of bees, wasps, and hornets of all kinds, is so nearly, if not quite identical, that I shall make no distinction between them. There are very few, if any persons, who do not know the symptoms, at least the local effects of the bee sting ; pungent, stinging, aching pain, redness and swelling of the part. The wound has, at first, and for some time, a white spot or point where the sting entered, surrounded by an areola of bright scarlet, growing fainter and paler as it recedes. The swelling is not pointed, but a rounded elevation, with a feeling of hardness. If upon the face, it not unfrequently causes the whole face to swell so as to nearly if not entirely close the eyes. In some instances, the brain becomes affected and death ensues.

TREATMENT.—I have for many years used but *one remedy*, and that has in all cases, and under all circumstances, when applied at any stage of the. affection, produced prompt and perfect relief, therefore I shall recommend no other. It is the common garden *Onion*, (*Allium cepa*) applied to the spot where the sting entered. I cut the fresh onion and apply the raw surface to the spot, changing it for a fresh piece every ten to fifteen minutes, until the pain and swelling, and all disagreeable symptoms disappear. If it is applied immediately after the stinging, the first application will afford perfect relief in a few minutes, and no further effect from it will be experienced. Applied later, it must be continued longer, and this may be done one or two days after the stinging, with just as much certainty of removing whatever symptoms may still exist. I treated one case when three days had elapsed, the patient (a young lady) was delirious and speechless, the whole face was so swollen as to entirely disfigure her features, raising the cheeks to a level with the nose, and closing the eyes. Her life was almost despaired of. The surface of a freshly cut onion was applied to the point where the sting entered, and changed about once an hour for a fresh piece. In a few hours consciousness returned, and a rapid recovery followed. All the swelling and disagreeable symptoms were gone in three days. *Ledum* is highly recommended by some physicians, and is doubtless of some value, but it is not to be compared with the *Allium*.

Bite of the Rattlesnake.—The most potent and certain remedy for the poison is *Alcohol*, in the ordinary form, or in common Whisky. Brandy, Rum or Gin. Let the patient drink it freely, a gill or more at a time, once in fifteen to twenty minutes, until some symptoms of intoxication are experienced, then cease using it. The cure will be complete as. soon as enough has been taken to produce even slight symptoms of intoxication. It is remarkable how much alcohol a patient suffering from the poison of the Rattlesnake will bear. An intelligent medical friend of mine in Kanawha County, Virginia, gave a little girl of ten years, who had been bitten by a Rattlesnake, over three quarts of good strong Whisky, in less than a day, when but slight symptoms of intoxication were produced, and that seemed to arise entirely from the last drink. She recovered from the intoxication in a few hours, and suffered no more from the poison of the serpent. Instances of cures with whisky are numerous, and I have never heard of a failure, when it was used as here directed. I presume it will do the same for the poison of other serpents.

Headache.—This symptom or affection, (if it can be classed as a disease) may depend upon so many causes, and be so very different in its effects, degrees of intensity, and the kind of pain or sensation attending it, that one will find it very difficult to mark out any definite treatment. I shall, therefore, only point out some of the more frequent cases, and the indications for certain remedies. What is called " SICK HEADACHE," or "*nervous headache*," begins by a sense of blindness or blur, before the eyes, of green or purple colors, dazzling or swimming in the head, without, for some time at first, any positive aching or pain. In the course of an hour, a longer or shorter time, the dimness of vision goes off, and the head begins to ache. This may or may not be accompanied with nausea and vomiting. Some persons are always more or less sick at the stomach, when these "nervous headaches" come on, others are not thus affected.

TREATMENT.—If taken as soon as the first blur before the eyes is noticed, or before any pain is felt in the head, *Nux-vomica* will, in nearly all cases, arrest the disease at once. It may be necessary to take two or three doses at intervals of an hour. Later in the case, though *Nux* may palliate, it will not cure. If headache with sickness comes on, *Macrotys* and *Podophyllum* should be given in alternation, every half hour, if the symptoms are very severe, and the nausea great; but in a mild case, give it once an hour, lengthening the interval as the symptoms abate. If the feet are cold, as is often the case, putting them into hot water will palliate the symptoms, and not interfere with the medicines. If the head feels hot, apply *hot* water to it. Never apply cold to the head, when there are any symptoms of congestion, as of fullness of the blood-vessels. For

Common Headache — If feverish give Gelseminum. If the face is red, and the arteries of the neck and temples throb violently, give *Bell.* If there is paleness and faintness, *Pulsatilla* is the remedy, especially if the forehead is principally affected. If the pain is mostly in the back of the head, *Nux* is to be used; if in the front, and is sharp, affecting the eyes, *Aconite;* if at the angles of the forehead, with a sense of pinching, *Arnica;* if a sense of fullness and pressing outwards, or with an enlarged feeling, *Macrotys* if intermitting or remitting, *Mercurius;* if there is ringing in the ears. *China.* Headache from fright should have *Aconite.* For that kind of *headache* that often occurs during the prevalence of fevers, and is not unfrequently a premonitory symptom of an attack of fever, I have found *Baptisia* and *Podophyllum* to be specifics. I give them alternately, every two hours a dose. until the headache ceases. It often subsides in a few minutes after the first dose of either. though I have sometimes failed with one alone and succeeded in the same cases afterwards with both in alternation. *I have* *no doubt* but that they act in many cases, as *prophylactics*, entirely warding off and preventing fevers, or at least arresting them at the premonitory stage. *Podophyllum* and *Macrotys* are most valuable remedies for headache.

Nose-Bleed—Epistaxis.—If it arises from fulness of the vessels of the head, with throbbing of the temples, redness of the face and eyes, *Belladonna* is the remedy. If fever is present, *Aconite* must be alternated with *Bell.* In females or children who have habitual nose-bleed, *Pulsatilla* and *Podophyllum* are to be used alternately, night and morning. During the paroxysm of bleeding. *Arnica* should be used, one dose repeated in a half hour if it continues. If it is produced by over-exertion, *Rhus* is the proper remedy. If it occurs in the *early stage* of fever, *Aconite* and *Bell.*; in the latter stage, *Rhus* and *Phos.* are to be used. *Hamamelis* will frequently arrest nose-bleed *immediately* after one or two doses.

Worms.—It is difficult to determine the presence of *worms* in children, much more in adults, yet both are affected by them occasionally. In children, there is more or less fever and restlessness, screaming out in sleep, starting, pain in the bowels, vomiting, choking, diarrhœa, picking at the nose, fetid breath, voracious and variable appetite.

TREATMENT.—*Santonine* is a remedy which

I have used for years, and I have treated many hundreds of cases, with such invariable success, that I feel disinclined to use or to recommend any other. It brings away the worms entire, and relieves the patient of all morbid symptoms immediately, or in much less time than any other remedy of which I have a r knowledge. It seems to act specifically upon the worms, causing them to leave the bowels by being evacuated with the fæces, without producing any sensible impression upon the bowels, the evacuations remaining natural, if they were so, or becoming so, if deranged, and the worms coming away not quite lifeless. I have often prescribed this remedy for children suffering under intermittent or remitting, and even typhoid fever, in the summer season, when there were not present any well-defined symptoms of worms, and yet the fever would soon abate, and in due time worms appear in the fecal evacuations. It often arrests entirely intermittent fever when worms are present, and are the probable cause of the fever. I give either the crude salt in from one-fourth to one-half grain doses, or first decimal trituration, from one to two grains of the trituration. Give one dose at bed-time, or in an urgent case at any other time, but never repeat the dose under thirty-six hours, and in an ordinary case, under forty eight hours.

This is *the* medicine *par excellence* for worms. It may be repeated once a week, when there is a tendency in the patient to the development of worm symptoms, or, in other words, the breeding of worms. The idea held out by some that it is hurtful, or unimportant to remove the worms, in itself considered, is simply *nonsense* and *worse*, for children are sometimes sacrificed to this idea.

Earacne—Otalgia —This may arise from various causes, but a common one is sudden cold. If it arises from cold, and there is general fever, or if the ear is red, or the side of the head and ear hot, *Bell.* and *Baptisia* should be given in alternation, every hour, or in a violent case, more frequently. These remedies will soon relieve such cases. Cloths wrung out of hot water should be laid over the ear, or the side of the head steamed, or it may be laid into water quite warm, with good effect.

Where the disease is a chronic affection, and the patient is subject to frequent attacks of pain in the ear, especially on a change of the weather, from dry to moist, *Mercurius* is the proper remedy, especially if it is worse at night, when warm in bed. If it arises from a shock or blow, *Arn.* is to be used. In scrofulous persons, whether there is ulceration or not, *Phosphorus* and *Pulsatilla* are the remedies. Children and even adults, not unfrequently suffer from earache, without any known cause sufficient to account for it. On examination into the ear you will often find either the cavity filled or nearly so, with a hard black substance, (the inspissated "ear-wax") almost as hard as horn, or else the ear will be quite empty, and the sides of the cavity *dry* and red, though perhaps not properly in a state of inflammation. The natural condition of the cavity as it can be seen by straining the ear outwards and backwards a little in a strong sun light, is moist, the surface covered slightly with a yellowish, greasy, soft sub-stance (the cerumen) "earwax." When this is wanting or in excess, or its character changed, it is evidence of disease, and pain is likely to occur.

TREATMENT. — Remove the accumulation when that exists, as the first step. But this must be first softened by pouring some warm oil, pure olive oil, or good pure sperm oil, into the ear, and repeat it two or three times a day for several days, until it is so far softened as to be easily removed with the probe end of common small tweezers, having a spoon-bowl point. When there· is dryness, moisten the surface with oil. In either case, it is best, for a while, to protect the delicate surface from the air, by putting oiled wool into the external ear. If the ear was filled, give *Mercurius* once a day until there appears a natural se-cretion. If dry, use *Belladonna.*

Toothache—It is difficult to determine the cause of toothache, and more difficult to select the remedy. · It often depends upon decay of the tooth, and exposure of the nerve to air, and contact with food or drinks, or even saliva, which irritate and produce pain.

Pulsatilla will as often relieve such cases as any other remedy, yet if it has been aggravated by a recent cold, *Bell.* and *Nux-v.* may be better. If the nerve is not exposed, and there is a disposition to a return of the pain on exposure to cold air, or a change of weather, the pain being of a *rheumatic* character, give *Rhus* and *Cimicifuga* in alternation. These will relieve many cases. For decayed teeth, the pain being dull aching with soreness, use *Chamomilla.* The body of the tooth, that is the dentine, sometimes becomes very sensitive when there is no decay or cavity, the pain being experienced when some hard substance hits, or the air or water, either cold or hot, comes in contact with the tooth. The temporary pain will generally yield to *Arnica*, and in most instances, the daily use of *Arnica* at the first decimal dilution, applied to the surface, and upon the jaws, will effect a cure. The *chloride of Zinc* applied to the surface of such teeth for a few moments will destroy the sensitiveness of the dentine. Teeth that are ulcerated at the roots, or have ulcerated gums around them, the teeth being decayed, should be extracted at once, for, besides the pain and inconvenience they cause, they are a *very prolific* source of *disturbance* to the digestive organs, from the positive poison generated by the decaying process. If people will use soft brushes upon the teeth with soap and water, followed by rinsing with simple water only, after each meal, brushing both inside and out and crossways, so as to clean between them, they will be saved much pain and decay, and disease of other parts, arising from foul and diseased teeth.

Teething of Children.—Affections arising from teething of children, are often of a serious character. The most prominent of which is *Diarrhœa. Fever* frequently accompanies the diarrhœa, and *convulsions* occasionally occur. *Aconite* and *Chamomilla* should be used in alternation, every one or two hours, according to the violence of the fever, and if convulsions occur, or are threatened, as will be known by twitching, starting, and screaming, use *Nux* and *Bell.* These may be given in rotation with the others, following the remedies, one after the other, every hour. I have relieved the most alarming cases in a day by this method of procedure. that had not yielded to either of the single remedies for several days, given as directed in the books; the patient growing worse continually. If the gums over the teeth look white, and the teeth, (one or more,) are near the surface. the gums should, by all means, be cut. Press the point of a lancet or penknife down upon the top of the gum, until the tooth is plainly felt, and be sure to make the cut as wide as the tooth. Rub the gums with *Arnicated water* once or twice a day. *Pulsatilla* should be given at night and *Chamomilla* in the morning, during the whole summer while the child is teething, as a prophylactic against the fever and diarrhœa that is likely to occur. It will generally save all trouble.

If the diarrhœa is profuse, watery and light colored or brown, give *Phos.-acid* and *Veratrum* alternately, as often as the discharges occur. For the restlessness of infants at night, *Coffea* is the specific.

Apthæ · Thrush —This is a disease peculiar to nursing children. The mouth becomes sore, and the tongue, lips, and fauces are covered with a white crust, looking like milk curds, which, when removed, leaves the surface red, inflamed and very tender. It sooner or later, extends to the stomach and bowels, producing severe and dangerous diarrhœa.

TREATMENT.—Of all the medicines known, none, according to my experience, will in the least, compare with the *Eupatorium·aromaticum*. It is almost certain to relieve speedily in all cases. I say this, not only from my own experience and observation, but from the testimony of several other homœopathic physicians, who have used it. It should be given at the first or second dilution, once in four or six hours, and three or four drops of the tincture put into a teaspoonful of water. and the mouth occasionally washed with the mixture. In summer, where agues prevail, and the child is feverish and restless. *China* will aid in the cure, to be given once in six hours between the doses of the *Eupatorium*. If the diarrhœa is obstinate, the discharges colored, and the child is sick at the stomach, give *Podophyllum* with the other remedies.

Inflammation of the Eyes—Ophthalmia. —For common Ophthalmia, in the early stages, while there is more or less fever and headache, with flushed face, bloodshot eyes and throbbing of the temporal arteries, *Bell.* and *Aconite* should be used alternately every two hours, and a wash made with ten drops of tincture of *Aconite* to one gill of pure water, applied to the eyes as hot as the patient can bear. This application should be repeated every two hours, in a violent case, until the eyes are easy, and then about twice a day until all inflammation and redness pass off. This will relieve a large proportion of cases in from one to four days. If, however, the case continues obstinate for a longer time, or has been of a week or more standing before the treatment is commenced, in the place of Bell., or after using it one or two days, use *Hydrastis* with the *Aconite*, giving them alternately at intervals of two to six hours, according to the stage of the case —more frequently as the symptoms are more urgent, using washes prepared of each separately, as directed for Aconite, except that the Hydrastis-wash may be twice as strong; and apply each about half as often as the same medicine is taken internally. The wash should, in all cases of acute inflammation of the eyes, be as hot as it can be borne. Let it be put into the eyes so as to come directly in contact with the inflamed surface. Simple hot water applied to inflamed eyes for hours together. allowing short intervals' between the applications, will often cure most painful cases. *Never apply cold* to inflamed eyes. It always aggravates. When the inflammation is in a scrofulous person, especially in infants, it assumes a purulent character, and may leave the cornea in clouded (nebulous) condition, and the sight more or less obliterated. For this condition use *Conium* first, and apply it *in tinct.*, half water, to the eyes every four hours.

Wounds and Bruises.—On this subject, I must necessarily be very brief. When a wound is inflicted, the first and most important thing to be done is to *arrest the flow of blood*. Every one should know how to do this. The bleeding is to be stopped, and the wounded vessels to be secured, so that no further flow can take place. First, then, to stop the bleeding, *pressure* is to be made upon the artery leading to the wound. If the wound is in the leg or foot, pressure is to be made, either on the vessel above and near the wound, or, where that cannot be easily found and compressed, make firm pressure with the thumb or some hard substance, in the groin,

about two and a half inches at one side of the
centre of the pelvis, (wounded side) just below
the lower margin of the belly, towards the
inner side of the thigh, where the great artery
(femoral artery) can be felt pulsating. By
pressing firmly upon this artery, the blood is
arrested in its flow into the limb, and of course
the bleeding from the wound soon ceases. If
the wound is in the arm or hand, *pressure* is
to be made, either just above the wound, or
on the inside of the arm, about one-third
of the way from the shoulder to the elbow,
where the artery (brachial) can be felt. To
secure the parts from further bleeding, the
wounded artery must be taken up and tied.
Let it be seized by forceps, or the point of a
needle may be thrust into it, and the vessel
stretched out a little, a thread put round it
and tied; cut off one end of the tie, and let
the other hang out of the wound, until it
comes out by the vessel sloughing off. Bring
the lips of the wound together, and if it is
large, put in stitches enough to hold them,
and put on an adhesive plaster, compress of
cloths, and bandages to keep it from straining
the stitches, and protect it from the air. · The
Arnica plaster, Sold by DR. LODGE, of Detroit, Mich.. is the best adhesive plaster of
which I have any knowledge. Give the
patient *Aconite* once in two hours, for a day
after the accident.

SLIGHT CUTS about the joints, especially

the knee, are dangerous. from their liability to affect the ligaments, inflame, and produce *Lockjaw.* Therefore, such wounds, ever so slight, are of great importance. They should be at once closed up, whether they bleed or not, and covered with an adhesive plaster, (Arnica plaster is the best) a bandage, and the knee should not be bent, even when walking or sitting, until the wound is healed. It is best to apply a splint from the hip to the heel, and bandage the limb to it, so as to prevent bending of the joint.

BRUISES are to be treated with *Arnica,* applied to the part affected, by putting twenty drops of the tincture into a gill of water. if the skin is *not* ruptured, or three drops into the same if it is, and bathing freely. The *Arnica* is to be taken internally at a higher dilution. Keep the parts covered with cloths and wet in *Arnica* water.

If a blow is received upon the head, by a fall, or in any other way, producing a "stunning" effect, (concussion of the brain) so that the patient appears lifeless for a time, and delirious when be begins to come to. there is great danger of inflammation of the brain, · and death from the re-action, or in some cases. the shock is so great that the patient will never revive unless he has the proper aid. *Arnica* is the great remedy to bring on reaction, arouse the patient, and prevent *dangerous* inflammation or congestion of the

6

brain   When a patient is "stunned" by a blow or fall, he should be conveyed as soon as possible, to some *quiet* place, and as little noise as practicable made about him, and the room kept darkened. *Arnica* 3d should be given immediately, and the nostrils wet with strongly arnicated water. If fever arise after he comes to, *Aconite* should be given with *Arnica* and if the head aches, or becomes hot, *Bell.* is to be used. This will prevent or arrest all symptoms of inflammation.

Torn and mangled wounds should not be handled much. If they bleed, the blood must be stopped as in any other case. If they are dirty, warm water may be gently applied to cleanse them. The wound should be covered with some soft cloths, and kept constantly wet in Arnicated water of the strength of four drops of the *tincture* to a pint of water.

Piles—Hæmorrhoids. — One important matter in all cases of habitual piles, is, to keep the bowels regular. Much can be done for this purpose by diet and regimen. On rising from bed in the morning drink freely, from a gill to half a pint of cold water, at least half an hour before breakfast; use such diet as is easily digested, and drink no alcoholic beverages.   To relieve the bowels when costive, take a dose of *Nux-vomica* at night, and *Podophyllum* in the morning. This may be repeated from day to day until the proper effect is produced.   To relieve from a severe attack of

Piles, use *Bell.* and *Podophyllum* in alternation
every four hours, and apply to the tumors
when inflamed, cloths wrung out of hot water,
or sit in hot water for a time. A poultice made
of fine-cut *Tobacco*, wet in hot water, and
crowded firmly up against the pile-tumors,
secured by a T bandage, will relieve the most
desperate cases for the time. and is attended
with no danger or disagreeable symptoms except in rare cases, when it produces sickness
at the stomach, which soon subsides on the
poultice being removed. *Oil of Arnica* is an
excellent application for inflamed Piles. A
most important point in the management of
Piles, and one often neglected, is to replace
the prolapsed tumors. The tumors will be
protruded from within the anus by the act of
evacuating, and if left in that condition, will
be pressed upon by the external parts, chafed
and inflamed. In all such cases, the patient
should take particular pains to return the
tumors into the rectum; and to aid in that
process a little oil may be applied when they
will be easily pushed back. and the sphincter
of the bowel will close below them, preventing
any chafing, and the consequent inflammation.

Bleeding Piles.—*Ipecac.* and *Bell.* are very
efficient remedies. They may be alternated
every half hour, or oftener if the bleeding is
severe, or at longer intervals when it is only
slight. *Hamamelis-v.*, will in nearly all cases
arrest the bleeding at once. It should be ap-

plied to the parts and taken internally at the same time. Drop doses to be put on the tongue once in fifteen or twenty minutes. An infusion of the *Hamamelis* may be taken internally in doses of half a teaspoonful, and the same injected into the bowels with excellent effect. The most effectual way, and the best for obtaining permanent relief from Piles when the tumors have become hard, and remain all the time so as to pass out of the anus at every evacuation, being constantly more or less tender and painful, and often becoming inflamed. is to have them taken off. But never let that be done with a knife. The bleeding would, in such a case, be very excessive, and most likely fatal. The history of knife operation for the excision of Pile tumors is written in blood, and the tombstone stands as a monument of condemnation of the practice. No trustworthy surgeon will at this day attempt it. But however dangerous may be the knife operation, there is no danger at all to be apprehended from removing the tumors by a *ligature.* To accomplish this, take a soft cork about three-fourths of an inch in diameter, and one inch long—make a hole through the centre from end to end, about one-eighth of an inch in diameter—cut crucial grooves in the top of the cork about an eighth of an inch deep, bevel down the lower end nearly to an edge, make a cord of saddler's silk, three fold twisted together and waxed, about eight or

ten inches long, double this in the middle and
pass the loop down through the cork out at
the sharp end. the two loose ends of the string
being out at the grooved end. Make a strong
hickory stick about three-sixteenths of an inch
in diameter, and just long enough to pass
across the square en l of the cork. Now have
the patient protrude the Pile tumors as far
out as possible, being placed on his knees with
the head bent to the floor, pressing out firmly ·
as if to evacuate the bowels. Let the tumors
be dried as much as possible by gently pressing a soft, dry cloth to them : then let the
loop of the string projecting from the flattened
end of the cork, be pushed on over the largest
tumor, and held down at its base, while an
assistant places the stick in one of the grooves,
ties the two ends of the cord firmly down
over the stick, or *toggle*, by a square bow-knot;
then turn the stick round once, twice or more,
until the pressure upon the tumor is sufficient
to strangulate it perfectly, and prevent the
string from slipping off. Care should be taken
to keep the cord down to the base of the tumor
while it is being tied and tightened. as in many
cases the base is much the larger part of the
tumor. and the cord tends to slip up. After
the ligature is applied and tightened. apply
arnicated water to the parts, and a large. warm
poultice of superfine slippery elm bark, wet
so as not to be too soft and slippery, on the
face of which Arnica may be put. Keep it

on with a T bandage. The patient must be put to bed and kept quiet until the ligature and tumor come off, which will be in about six or seven days, sometimes sooner. Once a day the "toggle" must be turned part, or the whole of a circle or more, to tighten the cord as the patient can bear. This will be very painful from beginning to end of the ligating, but any, even the most sensitive, patient can bear it. The patient must have quite warm hip-baths two, three, or more, times a day, or as often as the pain is severe, the poultice being replaced after each bath, and kept constantly on. If there are several tumors protruding, apply ligatures to two of the largest, when these are removed, the others will disappear. Injections of mucillage of slippery elm should be carefully used to move the bowels daily, or at least once in two days. Let the diet be of corn or oat meal mush, or rice. As the tumor gradually sloughs off, the surface heals, so that, though the base where the ligature was applied, may have been an inch or more across it, there will not be a raw surface of over an eighth of an inch in diameter, to which *Calendula Cerate* should be applied. The patient must keep quiet for a few days longer. Though this is a painful operation, it is not in the slightest degree dangerous. I have effected complete and permanent cures by this mode in numerous instances.

Sea-Sickness.– *Nux-vomica* should be used once in about four hours, for twelve hours before sailing, as a preventive to sea-sickness. If, however, symptoms, such as dizziness or blur before the eyes, and headache, begin to come on, a dose of *Nux* should be taken, followed in an hour with *Pulsatilla.* If the nausea comes on, *Ipecac.* and *Arsenicum* should be taken alternately between the paroxysms of vomiting, should that symptom appear. If practicable, the patient should lay still upon the back until the sickness passes off. I have removed sea-sickness immediately in several instances with *Pulsatilla* alone, and the last time I had an opportunity to prescribe for this affection I gave *Podophyllum.* It removed all the symptoms in a few minutes. That is the only time I ever tried it, but from the provings I am satisfied it is one of the best remedies.

Asiatic Cholera.—I was practicing in Cincinnati during the prevalence of Cholera in the year 1849, and 1850, and in Northern Ohio in 1854, and had abundant opportunity to observe and treat it. The disease generally begins with a diarrhœa, which may continue for several days, or only a few hours before other symptoms set in, such as vomiting, then cramping in the stomach and muscles of the legs, arms, hands and feet, followed by cold sweats, great prostration, restlessness, excessive and burning thirst, drinks being immediately rejected. These symptoms continue, the patient sinking rapidly into *collapse*, when the skin looks blue and shriveled, the eyes sunken, the surface covered with a cold, clammy sweat, the extremities, nose, ears, tongue and breath cold, the voice hollow and unnatural. This condition continues from two to eight or ten hours, the patient regularly failing, sometimes becoming delirious before he dies.

In some cases the vomiting and diarrhœa set in simultaneously, and the other symptoms follow, as above described, in rapid succession. In others the cramping may be the first symptom, the others following it.

In a large proportion of cases, the disease takes the course first described above, the diarrhœa, called the *premonitory symptoms*, or sometimes *Cholerine*, coming on several hours, if not a day or more, before any other symptoms.

The diarrhœa is not usually painful, hence the patient may not be alarmed so as to attend to it until the more dangerous symptoms appear. It begins in some cases with pain and some griping, the discharges rather consistent, having a bilious appearance, so that the patient supposes it to be an ordinary bilious diarrhœa, which is not dangerous, his fears being thus quieted. But however the diarrhœa begins, it becomes sooner or later, copious, watery, and light colored, (rice water) painless but rapidly prostrating.

TREATMENT.—In the early stages of the diarrhœa, *Veratrum*, taken about twice as often as the evacuations occur, will frequently arrest it in a few hours, especially if the patient lies down and keeps quiet. But if not, and it increases in frequency, or becomes more copious, or any sickness is felt at the stomach, the patient should, at once, be laid upon a bed and *strong tincture of Camphor* should be given in drop doses, once in five minutes, for one hour or more, and as the symptoms abate, once in ten, fifteen or twenty minutes, for six or eight hours. A teaspoonful of the *Camphor tincture* may be put into a tumbler of cold water. ice water if at hand, and the water agitated until it becomes clear, giving a tea-spoonful of this camphorated *cold* water as a dose, stirring the water each time. I think this is better than to give the pure tincture. After the patient becomes quiet and easy, *Veratrum* should be given in alternation with Camphor, a dose in four to six hours for several days, or oftener if he feels any symptoms like a threatened return of the disease. These two medicines serve as *prophylactics* (preventives) of Cholera. If, however, the disease continues in spite of the Camphor and Veratrum. in the first instance, or later, (as the Camphor may be given in many cases with success in the advance stage,) you must resort to other remedies. If vomiting comes on with burning in the stomach give *Ipecac.* and *Ar-*

senicum in alternation as often as the vomiting
occurs, and if the diarrhœa continues give
Veratrum between the doses of the other two,
in a violent case, as often as every ten to fifteen minutes, and at longer intervals when
the disease is slow in its progress. If the
vomiting and diarrhœa, or either, occur with
a kind of explosion, the vomiting ceasing suddenly for the time, after the first *gush*, or the
discharges from the bowels are involuntary,
Secale is the specific remedy. For the cramping, *Cuprum* and *Veratrum* are the remedies
to be given alternately. If, however, the
cramping comes on as the first symptom,
which is sometimes the case, the patient being
suddenly seized with it before any other
alarming symptoms occur, *Camphor is the
great remedy*, and in this case it may be given
in doses of double or treble the quantity before directed. If he sinks into the *collapse*
and lies quiet, indifferent to everything, the
pulse sinking, or he is pulseless. *Carbo-veg.*
will sometimes arouse and restore him, hope-
less as the case appears. It should be given
once in half an hour until the pulse begins to
rise. If, however, instead of being quiet he
is restless and thirsty, give *Arsenicum* in
alternation with *Carbo-veg.*, repeating the
dose as above directed. In some cases, after
all the active symptoms cease, the patient will
become quiet and drop to sleep, and instead of
the pulse rising, as it will if he is recovering,

it sinks, or does not appear if he has been pulseless. and the breathing becomes irregular and feeble—he. is sinking. If aroused, he sinks back into the stupor in a few moments as before. (*Laurocerasus* is a specific for this condition. It should be given once an hour until he is aroused.) If, however, besides the stupor, the head is ·hot, the face red, the breathing oppressed, the pulse slow and sluggish, *Opium* is to be used, and may be given in alternation with *Laurocerasus*. For the irritation of the brain. and furious delirium that sometimes sets in after the cessation of cholera symptoms, *Secale* and *Belladonna* in alternation will prove specific. Let the patient have warm or cold drink as he prefers, and let his covering be light or plentiful as is most agreeable. As soon as he gets easy, and the vomiting and purging cease, and his pulse begins to return, keep him as quiet as possible, let the room be darkened and everything still, so that he may go to sleep, which he is inclined to do, this being the surest restorer. 1 am quite sure I have known several patients carried off ·by a return of the disease, after it had been effectually arrested, in consequence of sleep being prevented by the rejoicing officiousness and congratulations of friends, disturbing and preventing that early and quiet slumber which nature so much needs, and must have, or hopelessly sink. The diet for two or three days after recovery, should be a little oat-meal gruel or rice.

Small Pox—Variola.—This disease begins with pain in the head and back, chilly sensations, followed by a high fever, so similar in all respects to a severe attack of bilious or "winter" fever, that it is difficult or impossible to distinguish it with certainty, as small pox. The fact of the prevalence of the disease at the time, and the exposure of the patient, may lead the physician and friends to suspect small pox. There is one very striking symptom of small pox, however, that exists from the beginning, which, though it may be present in fever simply. is not uniformly so. This is a severe and constant aching *pain in the small of the back.* The headache is also constant. The small pox is of two varieties or degrees, *distinct* and *confluent.* The *distinct* is when the pustules are separated from each other, each one a distinct elevation, with more or less space between them not affected by the eruption. The *confluent* is where the pustules spread out from their sides and run together, covering the whole surface as one sore. It may be distinct on some parts, as on the body, and confluent on others, as the arms, face, and parts most exposed to the air. In the *distinct* variety the fever continues without abatement until the eruption appears, when it entirely subsides, and that quite suddenly. The eruption comes out about the third day

of the attack, sometimes not discoverable until
the end of the third or beginning of the fourth
day. The eruption is at first very slight, beginning with small red pimples on the forehead, upper part of the cheeks, neck and
upper part of the breast, extending by degrees
to the arms, and other parts of the body and
limbs. About the end of the fourth or fore-
part of the fifth day, the eruption is complete.
There is a symptom, not mentioned in the
books, which will often determine the disease
before the occurrence of any eruption. It is
the appearance of hard shot-like pimples, to
be *felt under the skin* in the palms of the
hands, while there is, as yet, no trace of
eruption to be seen upon the surface. On the
eighth or ninth day, the eruptions become
vesicular, have flattened tops, and contain a
limpid fluid. The parts continue to swell, the
eruptions to enlarge, and become filled with
purulent matter, having a dark color at the
top, up to about the fourteenth or fifteenth
day, when they begin to flat down, to dry up,
and some of the scabs become loose. At this
time, some fever arises, often quite severe,
with headache and other inflammatory symptoms. If the eruption is very severe, fever
will be of corresponding violence, and lighter
or wanting when the eruption is mild. This
fever rarely lasts more than twenty-four hours,
from which time the patient rapidly recovers.
In the *confluent* variety, all the symptoms are

more violent. the fever continuing after the eruption begins. The pustules burst early, and run into each other, covering nearly or quite the whole skin ; the surface swells and turns black or dark brown, the lungs are more or less irritated, producing cough, and not unfrequently the stomach is nauseated, and vomiting ensues. If the patient survives the irritation up to the fifteenth or sixteenth day, when the *secondary fever* sets in, he is liable to be taken off by an affection of the brain or lungs. during this fever. If he recovers, his whole surface, especially that part exposed to air. is deeply pitted.

TREATMENT.—As it is not often known for a certainty. in the early febrile stage, that it is the small pox, the treatment will 'be first adopted that would be proper for a like fever arising from other causes. But in all my observations in this disease, and they extend to several hundred cases, I have not found in a single instance, any of the ordinary fever remedies, such as *Aconite* and *Belladonna*, which would be applicable for such symptoms in an ordinary case, to do any good in small pox. They are directed. however, for these symptoms by the authorities, in the febrile stage of the small pox ; but I am quite sure they are not the proper remedies. From the great similarity, the almost absolute identity of small pox *headache* and *backache*, with the same symptoms.developed by the *Cimicifuga-r.*

as well as the nausea and restlessness produced by the drug. I was led several years ago to the conclusion that this, or the *Macrotin* was valuable in small pox. Not only so, but during the prevalence of small pox in Cincinnati, to an extraordinary degree in the winter of 1849—50. I treated about one hundred cases, including both sexes, and all ages, from infants a few weeks old, to very old persons, giving the *Cimicifuga-r.* to all, and had the good fortune to see *all* my patients recover. Since that time I have prescribed it for every case successfully. Having then, been entirely successful in so many cases, with this medicine, I am not inclined at this time to give any other the preference. I must admit, however, that though my patients all recovered, I was not able to greatly abridge the duration of the disease, nor to prevent the development of all the stages in their proper order, as is *claimed* by M. TESTE, for his use of *Mercurius-cor.* and *Causticum.* I was satisfied with so far modifying the symptoms, as to enable my patients to live through, and come *out well in the end.* I would then direct. if small pox is suspected, the patient having been exposed to contract it, or from the peculiarity of the symptoms. in the early stage, or when the disease is discovered after the eruption, to give *Cimicifugin* at the first trituration, in one grain doses, once in two hours, while the fever, headache and backache continue, after which,

during the whole course of the disease, give it
three times a day. This will prevent the
development of a dangerous secondary fever,
as well as irritation of the lungs, stomach or
bowels. In addition to this medicine I give
the patients daily, from half an ounce to two
ounces of *pure (unrancid) Olive oil.* This
serves to prevent the development of pustules
in the throat, lungs and stomach; is more or
less nutritious, and keeps the bowels in a
healthy condition. Wash the surface once a
day in weak soap suds, following it with a
bath of milk and water, and keep cloths
moistened with warm milk and water, constantly upon all parts that are exposed to the
air, lubricating the surface with *Olive-oil* after
the bath of milk and water. This keeps the
surface quite comfortable. The best diet is
corn or oat-meal mush and molasses, to be
taken in small quantities. Cold water is the
proper drink, though it should not be very
cold. The room should, at all times, be well
ventilated, but in cold or cool weather, sufficient fire must be kept up, to keep the room
warm and dry. A temperature of about 65°
is the best. Hardly any thing can be worse
for a small pox patient than to be in a cold or
damp room, and to breathe *cold* air. Uniform
temperature is importan . If the eruption is
tardy about appearing, or after it is out, a
recession takes place, the Alcoholic Vapor
bath will soon bring it out. (See Rheumatism.)

Occasionally the feet and limbs below the knees, will swell prodigiously, and become extremely painful, causing the principal suffering. For this. wrap the feet and legs in cloths wet in a strong solution of Epsom salts, quite warm, and cover with flannels so as to keep them warm. This will afford immediate relief. and reduce the swelling in a day or two. The finely pulverized Epsom salts, dry, sprinkled on the pustules. will very often prevent pitting. It is the safest and surest remedy of which I have any knowledge.

Varioloid is small pox modified by vaccination. It is to be treated as a mild case of small pox. The *Cimicifuga-r.* has been used with apparent success as a prophylactic (preventive) to small pox, taken three times daily.

Painful Urination. Incontinence of Urine.—INVOLUNTARY URINATION. — Where the discharge of urine produces smarting and burning of the urethra, *Cantharis* is the remedy. Where there seems to be an over-secretion of acrid urine, producing inflammation of the neck of the bladder, known by pain in the glans penis. *Copaiva,* and *Apis-mel.* are the remedies. If there appears to be a partial palsy of the neck of the bladder, the discharge taking place in sleep, *Po lophyllum* is the surest remedy. I have cured some bad cases by the use of these three remedies. given in rotation three or four hours apart. Injections of a solution of borax into the bladder, have, in several cases, been sufficient to effect a perfect cure, without any other remedy. This may be used in connection with the other remedies. [Nitrate of uranium is one of the most reliable medicines for incontinence of urine.]

For *painful urination* with a distressed feeling in the neck of the bladder, causing a constant disposition to evacuate urine, the *Althæa-officinalis* is a certain remedy ; it acts like a charm. It is an important remedy for inflammation of the bladder. A good mode of using it is in form of a warm infusion in doses of a table spoonful every half hour or hour. according to the urgency of the symptoms. The *Althæa-rosa* (Hollyhock) may be used as a substitute, though it is not as good. Every family should cultivate the *Althæa-officinalis* (Marsh-mallow), so that the fresh green root, which is the best, can be procured at any time. I have been able to relieve patients with it, especially females, when all other remedies seemed unavailing. It is particularly useful for urinary difficulties of pregnant females.

Neuralgia.—*Aconite, Belladonna* and *Gelseminum* are three important remedies in this affection. If given low, and applied directly along the course of the affected nerves, at full strength of the tincture, they will almost always effect a cure. The proper way to use them is to give them internally at the second dilution, at intervals of fifteen to thirty minutes, when the pain is severe and nearly constant, and apply *Aconite tincture* as hot as practicable over the course of the nerve, by means of wet cloths, for an hour or two hours, and if the pain has not subsided use *Bell.* locally in the same manner. If the Neuralgia is periodical, coming on at regular intervals, *Arsenicum* and *China* are the remedies. and they should be used externally as directed for the others. both at the first dilution, and given internal y at intervals. in proportion to the violence of the symptoms, the *Arsen.* at the 3d and the *China* at the first dilution. If the patient has used alcoholic drinks to excess, *Nux* is to le used in place of Arsenicum. *Periodical Neuralgia* generally requires the same treatment as ague. In females when there is uterine disease, *Pulsatilla* and *Macrotin* are the remedies to be used, as directed.

Jaundice — This disease depends upon derangement of the liver. The skin and whites of the eyes become yellow; the patient grows weak, loses his appetite, is dull and sluggish in all his actions, melancholy and discouraged in his moods.

TREATMENT. — *Mercurius* and *Podophyllum* given in alternation, each twice a day, will near'y always effect a cure. If the patient is costive. *Nux* should be taken at night, until his bowels become regular. Bathing the surface daily, or oftener, is a very important

measure in the treatment of this affection. As
often as once in two or three days, an alkaline
bath should be taken. If the patient has fever
every day, or once in two days, ever so slight,
China should be used with *Podophyllum.* If
he has been drugged with Mercury in any
form, in large doses, even six months or a year
before, give *Hydrastis* in place of Mercurius.

Itch. —I shall say but little about this very
common and very obstinate affection. - Every-
body has a " cure for itch," yet nobody cures
it short of the use of *Sulphur* in some form.
Though the attenuations of Sulphur may sometimes cure itch, it must be acknowledged that
such cures are so rare in this country, and the
time requisite to accomplish it is so long, as a
general rule, that few will trust them. The
most successful remedy, and the one that will
always cure quickly, if at all, is *Hepar-sulphur-kali* (Sulphuret of Potassa). To succeed
with it most certainly, let the patient be
thoroughly bathed with warm soap suds,
quite strong, in a room at the temperature of
90 to 100°, continuing the bathing and *rubbing*
for an hour or more, then dry off the surface
with soft cloths, and apply the *Hepar-sul.-k.*
with water, at the strength of thirty drops of
the strong alcoholic solution, with a gill of
water, wetting every eruption on the whole
surface and let it dry on. This causes some
smarting, but it is effectual; it kills the *acarus*,
(itch animalcule) and in a few days the sores

heal, the itching all subsides immediately. If every pustule has not been touched, those left may continue to itch, in which case, a second application is necessary. *Hepar Sul.-calc.* should be given internally at the third dilution, for a month, once a day, after the baths. Avoid greasy food. For the

Scald Head of children, where there is a discharge of yellow and watery pus from the sores, and the eruption extends to the ears or face, like the disease called the *crusta lactea* (milk crust) the same washes as for itch are the most effectual, while at the same time, and for a month, or two, the child should have *Hepar-sul.* 5th at night, and *Petroleum* 3d in the morning. Daily ablution of the head with warm soap suds, and keeping it covered, are absolutely essential.

Carbuncle.—This affection, though it somewhat resembles a common boil, and is by some writers considered only such, in an overgrown state, is, nevertheless, far from being identical with it. While a *boil* is only a sanitive effort of nature to eliminate the cause of a morbid process, and tends to a spontaneous, healthy termination, the *carbuncle*, on the contrary, is the very essence of disease; its constant tendency being towards the dissemination of diseased action, causing destruction of the parts affected. It, in fact, appears like a parasite, living by the destruction of surrounding tissues, literally absorbing them and "thriving on death." It begins with a red, livid color, slight aching and burning pains, the part swells and is elevated some like a boil; except that it does not "point." but has a broad base rising like a cone and flattened at the top. It feels soft and spongy, and will appear to fluctuate, but if punctured, blood only flows. The pain and burning increase rapidly, and sooner or later several openings appear upon the top, varying from three or four to half a dozen or more, looking like the holes in a sponge, out of which issues a fluid like thin gruel. Instead of becoming easier after the suppuration begins, as is the case with a boil, the burning increases to an alarming and unbearable extent; cold chills, loss of appetite, great depression of spirits, general nervous and muscular debility come on. The tumor continues to discharge, turns purple; gangrene beginning in the carbuncle extends to other parts and death follows. The disease is nearly always confined to quite feeble persons and those past the meridian of life; but I have seen it on younger though feeble patients. It is generally located on the back, occasionally on the head, where it is very dangerous from its liability to affect the brain.

TREATMENT.—If treated very early, *strong tincture of Arnica*, applied to the surface of the carbuncle, by cloths wet and laid over the tumor, will often arrest it so that the swelling will not be developed to the suppurative stage. However, to reap any benefit from *Arnica*, it must be applied while the pain is not severe, and the parts only feel bruised and tender to pressure, like a common bruise. After the ulceration occurs. *Arsenicum* is the great remedy to be relied on. It should be given at the second or third attenuation as often as every three hours, when the pain is severe, and applied to the surface of the carbuncle freely by cloths laid over it, wet in the first dilution, or by sprinkling the first trituration of the oxyde (1–10) freely upon the open surfaces, so that it may penetrate into the open mouths or orifices. Over this powder apply an emolient poultice, or soft cloths wet in water hot as can be endured. This will soon allay or greatly lessen the pain. It should be repeated as often as any of the burning pain peculiar to the carbuncle returns, until the tumor suppurates in a tolerably healthy manner; then lessen the strength of the *Ars.* applications, and continue them until it has the appearance of a healthy abscess, when only simple dressings are necessary. Some may suppose such strong applications injurious, but I can assure them from abundant experience. that there is not the slightest danger. The carbuncle should *never be punctured* or *cut into*. Such operations always make them worse, and induce a more rapid approach to gangrene. The patient should have nourishing food.

Good native wine taken in moderate quantities, by a very feeble person, has been recommended. [Small quantities (5 drops) of pure tincture of China two or three times a day, on sugar, is preferable.]

Though the knife operations for the removal of carbuncle are always injurious, the chemical effect of *Potash* is frequently most beneficial. I have, in repeated instances, applied to the ulcerated surface. *caustic-potash* freely, allowing the dissolved caustic to penetrate to the very "core" by running into the orifices. At first it would produce some smarting. but the pain is different from that of the carbuncle, and the change is agreeable rather than otherwise. Soon after the application all pain ceases, and the tumor, under the use of a poultice, begins to slough off in a few days, leaving a raw surface, disposed to heal kindly, Occasionally, however, the healing process is tardy, when *Arsenicum*, at the third applied and taken internally, will soon effect a cure.

I have occasionally used *Hepar-sulphur-calc.* with good effect in the latter stage.

Fel n—Whitlow. —For this disease, in the early stage, when the sensation is that of sharp, sticking pain, feeling as though a brier or thistle was in the finger, immerse the part in water as hot as possible, into which put common sa t as long as it will dissolve; hold it in this *hot* salt bath for an hour or more at a time, and when removed, apply finely pulver-

ized salt, wet in *Spirits of Turpentine;* bind
on the salt with several thicknesses, and keep
it constantly wet with the Spts. Turpent. for
twenty-four hours, when, if all symptoms of
felon are gone, no further treatment is necessary. As a general rule, the hot bath should
be repeated three times a day, especially if the
symptoms have existed for several days and
there is much pain or swelling, and the
dressing should be kept on as above directed
for several days, more or less, until all symptoms disappear. I am quite confident that a
large majority, if not all, of the cases if thus
treated at any time before pus is formed, will
be discussed and cured. If pus has begun to
form before the treatment is commenced, this
will not *cure* the felon, but it is good treatment, especially the hot bath, as it will greatly
lessen the pain. By holding it in hot water
for an hour or two each day, the suppurative
process will be hastened, and as soon as the
pus can be felt at any point, fluctuating, puncture and let it out; then continue the hot
bath, with *Calendula (Marygold)* flowers in
the water, keeping the part all the time warm
and moist. For the restless and nervous
irritability that frequently occurs, especially
in females, *Aconite is the best remedy.* It
should be given, one drop of the tincture to a
gill of water, in teaspoonful doses, once in one
or two hours, and the same applied to the sore.

DISEASES OF FEMALES.

Suppression of the Menses, (Amenor-rhœa).—For sudden suppression from taking cold, as by wetting the feet, there being headache, more or less fever, the pulse frequent and variable, pains in the small of the back and cramp-like pains in the pelvic region, give, in alternation, *Aconite* and *Pulsatilla*, as often as every fifteen or twenty minutes in a violent case, and at longer intervals as the patient begins to get easy. Putting the feet into hot water, or taking a hot sitz-bath is very useful. If the patient is sick at the stomach, as is often the case, give luke-warm water freely and let her vomit; after which let her drink freely of water as hot as it can be safely swallowed, adding milk and sugar to make it palatable. The good effects that are often attributed to and experienced from the use of various hot teas in this affection, are, in my opinion, attributable more to the hot fluid alone than to any specific medicinal virtue in the substance of which tea is made. At all events, very *hot* drink with nothing but water, milk and sugar, is equally efficacious, and my medicine (a few grains of sugar of milk) put into the hot water, seasoned as above, has often obtained great credit, when the *hot water* was alone worthy. Rubbing the loins and abdomen briskly downwards

106

with the hands of a healthy and vigorous
nurse, will often excite the menstrual flow
after a sudden suppression. If the head is
hot, the face full and red, and the arteries
of the neck and temples beat violently, give
Bell. with *Pulsatilla*, and if the lungs are
oppressed, use also *Bryonia*, giving the three
in rotation. If, after the menstrual flow begins, there is still much pain in the pelvic
region, give *Caulophyllum*, which will immediately afford relief *Apis-mel.* is very
servicable in suppressed menses of several
days or even weeks duration, where there is
fever, redness of the face, and pain in the head,
and pains in the hips extending to the limbs,
especially if there is any tendency to bloating
of the abdomen and swelling of the limbs or
feet. It acts *promptly* and *efficiently*. If the
suppression has been caused by sudden fright
or any strong mental emotion, *Veratrum*
should be given in connection with the two
former medicines. Should there be great fullness of the vessels of the head, or bleeding at
the nose, *Bryonia* with *Pulsatilla* are to be
used. *Bell.* is also useful in this case if the
pain in the head is throbbing, especially if any
delirium is present. For suppression in young
females, of several months' duration, I have
used, with much success, *Podophyllum* and
Cimicifuga one at night, the other in the
morning, giving them for two or three weeks
before the proper time for a return, and a day

or two prior to the time, give also *Pulsatilla,* and give the three in rotation, a dose every six hours. This practice has been successful with me in cases of long standing and apparently obstinate character. Where there is other disease. as an affection of the liver, lungs or stomach, .this must be treated and cured, or the menses will not probably return. Great care should be exercised to keep the patient's feet and limbs warm, as upon this may depend her future health.

Dysmenorrhœa.—Painful Menstruation. —For this disorder, I know of no one remedy so valuable as the *Caulophyllum,* but *Pulsatilla* in many cases is efficacious. and as they do not prevent each other's action, I prescribe them in alternation, giving a dose every half hour, for a short time during the paroxysm, or until the pain abates to some extent, then every hour. If there is pain in the head, sickness at the stomach, a kind of sick headache, as is often the case, with painful menstruation, *Cimicifuga* should be used with the others; *Ipecac.* is the *specific* for an excessive flow of the menses with great pain, especially if the stomach is nauseated. It should be given as low as the first dilution, and the tincture, in water, in the proportion of thirty drops to half a pint, injected into the vagina quite warm. The application of extract of *Belladonna* to the neck of the uterus will often produce immediate and perfect relief. After the patient is relieved from the painful paroxysm, she should be treated so as to prevent a return of the pains at the next monthly period. *Pulsatilla*, *Caulophyllum* and *Podophyllum* are the three medicines that are most certain to effect this object. They are to be given one medicine each day, a dose at night for three weeks, then morning, noon and night, until the time for the return of the menses, when they should be used oftener if there is pain. If the patient is inclined to be costive, *Nux* should be given at night for a few days before the menstrual period, in place of *Pulsatilla*.

Menorrhagia — Profuse Menses – Flowing.—For this affection *Ipecac.* and *Hamamelis* are the specifics. They should be taken alternately, at intervals of from half an hour to two hours apart, according to the urgency of the symptoms, and the *Hamamelis* injected into the vagina. These will nearly always arrest the flooding immediately. *Secale* should be used either alone or with the above medicines, if there are bearing-down pains like labor pains, and sickness at the stomach in spite of the Ipecac. *Ipecac.* alone is often sufficient.

Nursing Sore Mouth. — Sore mouth of nursing women, as the name of the disease indicates, is peculiar to women who are suckling children. It is an inflammation of the mouth, tongue and fauces, which sometimes comes on during pregnancy, several months or

but a few days before the birth of the child.
It generally, however, makes its first appearance when the child is a few weeks old, and
sometimes not till after the lapse of several
months. In some cases the tongue and inside
of the mouth ulcerate, and the irritation extends to the stomach and bowels, producing
distressing and dangerous inflammation of
these parts, with severe and obstinate diarrhœa. For the sore mouth, before diarrhœa
begins, give *Eupatorium-aro.* and *Hydrastin,*
in alternation. a dose once in three hours, and
wash the mouth with the same. each time.
After the diarrhœa occurs, use *Podophyllum*
with the other medicines, giving them in
rotation, three hours apart. It is best to give
a dose of *Podophyllum* night and morning. I
have treated very bad cases of this disease
that had been running for more than a year,
and been treated with the ordinary remedies
directed in the homœopathic authorities without any permanent benefit, curing them perfectly in ten days with *Podophyllum* and *Leptandrin,* giving them in alternation at the 1st
attenuation in half grain doses, at intervals of
from four to eight hours according to the frequency of the evacuations. These two remedies
are almost certain to arrest *Chronic Dysentery*
where there is ulceration of the lower portion
of the rectum, a peculiar distress felt at the
stomach just before stool, with *sudden* rush of
the evacuations and inability to control the

inclination even for a few minutes, with a
feeling of faintness after the stool. *Leptandrin*
is the specific for the Dysentery that often
succeeds cholera, and these two, *Pod. and
Lept.*, are almost certain to relieve the
"Mexican Diarrhœa," as well as that connected with the fevers along the Mississippi
river.

Mammary Abscess.—(*Ague in the breast—
Inflamed breast.*)—This is a disease peculiar to
nursing women. The first symptom is a slight
pain or soreness in some part of the "breast,"
which continues to increase for a day or two,
when a chill, more or less severe, sets in, followed by high fever and quick pulse, headache and great restlessness. The gland swells
and becomes very painful. This is generally
a disease of rather slow progress, running
eight or ten days and sometimes two or three
weeks before abscess forms and "points" to
the surface.

TREATMENT. — *Phosphorus* is to be taken
internally, and the first dilution put in water,
twenty drops to one gill, and applied to the
surface by means of cloths wet in the mixture,
as hot as it can be borne, and laid over the
whole breast. If this is done and the medicine
given internally every hour, as early as the
first and frequently as late as the second or
third day, it is quite sure to remove the disease and prevent an abscess. It is best to use
it even much later. In fact it often succeeds

as late as the fifth or sixth day, and if it does
not prevent the abscess, it so far palliates the
severe symptoms as to render the pain but
slight and keep the patient comfortable. An
application of the tincture of Cantharides,
diluted with water and applied to the breast
by cloths wet in it, to the extent of producing
considerable redness and even eruptions, and
the second dilution of the same taken in drop
doses every three hours, has proved successful
in subduing the inflammation after *Phosphorus*
had failed, and it was supposed an abscess
would form in spite of any treatment. I recently succeeded in giving perfect relief with
Apis-mel. internally, applying it externally
after the pain and swelling was very great. I
am of opinion that the *Apis* is a valuable
remedy. *After abscess forms* as soon as the
pus can be felt at any point, soft and fluctuating under the skin, *puncture* and let it out,
then poultice it for a few days until it heals,
giving *Phosphorus* and applying it to the sore.
In *puncturing*, always be *very particular* to
have the lancet or knife enter so that the edge
will look towards the point of the nipple, so
as not to cut *across* the milk ducts, which all
run toward that point, and if cut off will close
up so that the milk which may be secreted at
any future time cannot get out, and swelling,
pain and severe inflammation, abscess and
ulceration will be the consequence; whereas,
if the cut is made lengthwise of the ducts,

very few, if any will be cut off, and all future danger will be avoided. Apply an elm poultice from the beginning to the end of treatment. For malignant ulcers of the breasts, the *Cornus-sericea* is a most potent remedy. It is to be taken internally at the first dilution, and applied in strong infusion or diluted *Tr.* of the bark to the sore.

Sore Nipples — This affection of nursing women frequently comes on before the birth of the child, but generally does not make its appearance until after the suckling has continued for a week or more. It seems in some cases to be connected, with the aphthæ (sore mouth) of the child, or at least to be aggravated by contact with the sore mouth; on the other hand it sometimes seems as though the sore nipples produced the sore mouth of the child.

TREATMENT. — I treat both the nipple and the child's mouth with the same remedy *Eupatorum-aro.*, applied at the strength of 6 drops of the tincture, to a teaspoonful of water, the application being made by a soft cloth, wet and laid over the nipple; give drop doses of the same strength internally every three hours, which will, in nearly all cases effect a cure in one or two days. The child's mouth should be wet with the same each-time just before nursing. The oil from the pit of the butternut. (Juglans-cinerea,) obtained by heating the pit and pressing out the oil, applied to the nipple, will generally cure it after 3 or 4 applications about six hours apart. The child may take hold when the oil is on, without danger. This remedy is sufficient in nearly all cases. [A very effectual application is the glycerole of *Hydrastis*.]

Leucorrhœa and Prolapsus Uteri — Whites. Female Weakness.—The disease depends in all cases upon *inflammation* of the uterus, or vagina, or both. The inflammation may be simply in the neck of the uterus extending to the posterior surface of the vagina, or the latter may not be affected; or it may extend to the whole internal surface of the uterus, producing swelling of that organ, both the fundus and neck. The swelling may be confined mostly to the fundus, causing it to be too large for the space it ordinarily fills, hence there will be more or less *displacement* of the womb, and crowding upon other parts, as the bladder or rectum. In some cases, the swelling is more on one side than on the other, so that it will be crowded over to the opposite side These displacements are often called *prolapsus uteri*, or "*falling of the womb*," carrying the idea that the difficulty depends upon a morbid relaxation of the ligaments that support the organ. Not one case in a hundred is of this latter character, but nearly, if not all, depend upon the inflammation and swelling above mentioned. How futile then, not to say *hurtful*, must be all instruments

for, and all·attempts at replacing and supporting it by *force!* All such mechanical
meddling is injurious, and should, with all the
" supporters," be condemned and discarded.
They may afford temporary relief, but this is
at the expense of future health. Cure the
disease, relieve the inflammation, and nature
will replace the organ. Leucorrhœa is always
present where there is ulceration of the neck
of the womb, and this ulcerated condition
e~sts to a greater or less extent, in many
cases where it is not suspected by the patient.
It is vastly more prevalent than is generally
supposed. The *symptoms* are numerous.
Among the more prominent are a sense of
weight and bearing down in the pelvis, pains
extending down the limbs, aching aud weakness of the small of the back, headache, more
or less gastric disturbance, dyspepsia, the food
souring on the stomach. There is often,
especially when there are ulcers on the parts,
a distressing sense of heat or a smarting
sensation The menstrual function is frequently deranged. the bowels costive, the
urethra, by being pressed, becomes irritable
and burns and smarts whenever the urine is
evacuated. The sleep is disturbed and un-
refreshing, and the whole nervous system is
unstrung. The discharge from the diseased
surface is an ordinary case without ulceration,
is of a mucous or muco-purulent character,
not unlike an ordinary catarrhal secretion.

When ulceration exists it is dark, fetid or bloody, or sanious and purulent, sometimes it is acrid, excoriating the parts.

TREATMENT. — Inflammation or ulceration, either acute or chronic, in these parts does not differ essential'y in its characteristics from the same affection in other mucous surfaces. The proper treatment for a catarrh of other mucous surfaces will be applicable to these, though there is no doubt but that some medicines are more specifically adapted to these than to other organs. In the early stage of the complaint, while the inflammation is acute, or sub-acute, the discharge thin or white, *Copaiva* and *Cimicifuga* are to be given once in 6 hours alternately. During the same time let injections into the vagina of warm soap and water be used twice a day, to cleanse the parts of the secretion, followed in half an hour by a wash of warm water, into which *tincture of Cimicifuga* has been put in proportion of 40 drops to half a pint. The application should be made with an 8 ounce or, at least 6 ounce curved pipe syringe, so as to throw it with considerable force. If there is a burning sensation, use the washes quite warm, until the heat of the parts is allayed. Avoid the use of *cold* injections as long as any inflammation exists If the bearing down is present with burning in the parts, *Bell.* is to be used in rotation with the two former remedies. If the sensation is that of smarting,

Cantharis is to be used in place of Bell. Where the disease comes on soon after child-birth, *Podophyllum is the specific.* It is to be given at the first attenuation three times daily in half gr. doses of the trituration. In this case let the parts be freely washed daily with a solution of borax, quite warm. In the *chronic* form of the disease, especially where *barrenness* exists, *Cimicifuga*, *Podophyllum* and *Hydrastis*, given morning, noon and night, in the order named, will, in nearly all cases, afford relief. For females who have never borne children, give *Phos.-acid*, 2d and *Eryngium-aquaticum* 1, night and morning for a week, and then give them at the 3d dilution until the symptoms subside. If there are headache and derangement of the stomach, *Cimicifuga* and *Podophyllum* should be used, each once a day, between the latter remedies. When the discharge is colored and the pains darting, cutting or smarting, indicating ulceration, or if ulceration is discovered by examination, use *Macrotys* and *Hydrastis* internally, injecting the latter upon the affected parts freely. The ulcerated surfaces should be well washed off every day with soap and water, or a solution of borax, and the medicine (*Hydrastis*) in form of infusion, used half an hour after the other wash. If the neck of the womb looks dark, and is ulcerated, or is hard and painful to the touch, especially on probing the cavity, *Cornus-sericea* must be used both as a wash to the parts, and at the first dilution internally, using them twice a day. This remedy will often cure malignant cases. It takes a long time in some instances to cure a chronic case. but if persevered in, these remedies will not be likely to fail.

(NOTE.—The late Prof. Morrow was remarkably successful, and became justly celebrated for curing hard cases of Leucorrhœa ulceration and "Prolapsus uteri." Almost his entire reliance in their treatment were the *Cimicifuga* and *Caulophyllum* given internally and by injection upon the parts. He gave the *Cimicifuga* in the form of tincture every day to the extent of producing specific head symptoms, when he discontinued it till the next day, using the Caulophyllum in the meantime in small doses. He rarely if ever failed.) .

Morning Sickness of Pregnant Females. —The most efficient and certain remedy for this symptom is *Macrotin*. It should be taken at the first attenuation, a dose before rising in the morning, and one every six hours during the day, as long as the sickness is troublesome. It will generally relieve in a few days. If the stomach is sour use *Pulsatilla* with the *Cimicifuga*. As a *preparation for labor*, a dose (one grain) of *Macrotys* at the first attenuation given in the morning, and the same of *Caulophyllum* at evening, is of great service.

Whatever others may think or say in relation to any preparatory treatment for labor, I have reason to know as well as anything in medicine can be known, that patients treated as here directed, pass through labor much quicker, frequently in one-fourth the usual time. Their sufferings are comparatively trifling, and the length of time for recovery to ordinary health, after labor, is abridged from three-fourths to nine-tenths that of former labors I am quite confident the medicines produced this difference. For *irregularity* of *labor pains*, and for distressing *after pains* the *Caulophyl'um* is specific. During labor it should be given at the 2d attenuation in about half grain doses, every half hour, until the pains are regular. Two or three doses at most, and generally one will suffice. For the after pains it may be given in alternation with *Ipecac.* or *Aconite* if there is flooding, or with *Pulsatilla* when the flooding is not troublesome, a dose once in half an hour, until the pains are checked. For *rigidity* of the soft parts and severe, *retarded and long protracted labor*, where the pains are strong or irregular, and great pain and exhaustion is experienced on account of the unyielding condition of the parts, *Lobelia-inflata* given in drop doses of the Tr. in water, once in twenty minutes, in alternation with *Caulophyllum* as above directed, will in a short time produce the proper condition of the parts, while they render the pains stronger, regular and progressive. In urgent cases I have given the medicines every 5 or 10 minutes, with decided benefit. [Gelseminum is doubtless preferable to Lobelia.]

A Useful Hint to Mothers. — Children push beans, peas, corn. &c., into the nose and ear, causing much alarm. To remove such a body take a syringe that works tightly, put the end of the pipe against the bean, shot, or other substance, draw back the piston so as to *suck* up the article firmly as the pipe is withdrawn from the cavity.

Local Applications —That medicines act locally, that is, manifest their symptoms by peculiar derangement or disturbance of some particular part of the system, more prominently than of any other part, for the time, no one will deny. That each one has some particular locality or tissue upon which its action is more perceptible than anywhere else, is equally undeniable, and that the prominent symptoms are often external and local, is also true. Yet, with these truths clearly demonstrated, there are those of our school who discard the external or local application of all remedies except *Arnica*. Why this is done, is difficult to determine, unless we can believe that such physicians suppose it to be *heresy* to make use of any remedy in a different manner from what was recommended by the "Father of Homœopathy," and abjure all possibility of *improvement* in our practice.

That nearly if not all medicines, may be applied externally with advantage, when there are local manifestations similar to those produced by the drugs, there can be no doubt in the mind of any sensible man. That they will act favorably when so used is *reasonable*, as a matter of theory, and that they do, as a matter of fact, has been *proven* to my mind, by abundant experience in their use. Therefore, I hesitate not to recommend the practice to others. Medicines must act either by combination with the affected part, or by *catalysis*, changing the molecular action of the living tissues. In either case, they must come directly in contact with the part to be affected. This *must* be done through the circulation, when taken internally, or it *may* be done by direct application of the remedy to the diseased tissue, when that is so situated as to be reached. The difference is greatly in favor of the latter mode when that is practicable, from the greater certainty of its results. This assertion is based, not upon vague hypothesis, but upon *actual practice*. Entertaining these views, however heretical they may be pronounced, I shall proceed to mention some of the remedies I have learned to use thus, and the cases for which they are prescribed. I would remark that, in selecting a remedy, it must be done with as much certainty of its homœopathic relation to the local or general symptoms for external as for internal use. I have found, however, that much lower attenuations are requisite and admissible.

ARNICA is highly applicable to *bruises*, and is valuable also when applied to lacerated or mangled surfaces. to the surface of the limb where a bone is fractured, also about the joint when it has been dislocated. It is to be used in the form of *Arnicated water*, by putting one or two drops to a gill of water for application where the skin is ruptured or the surface raw. and ten to twenty drops to the gill, upon parts where the skin is sound. It is useful also. for *boils*, and *carbuncles* in the *early stage*, the *strong tincture* to be applied when the surface is sound, and (to boils) when the surface is open. one drop to a gill of water.

ACONITE.—Is applicable to inflamed eyes, in the early stage, where the disease is in the conjunctiva, (that portion which lines the lids and covers the front of the ball), especially if there is a sense of scratching, as though some foreign substance is in the eye, great intolerance of light, chilly sensations, with more or less fever, and quick pulse. Put three or four drops to a gill of warm water, and apply it freely. It is also very valuable for *Neuralgia*, applied strong and warm, along the course, or at the origin of the affected nerve. In neuralgia of the face, apply it upon the side of the face, also just behind and below the ear of the affected side. It is of much value as a remedy for neuralgic affections of the womb. I have relieved the most distressing symptoms of neuralgia of the womb, in a few minutes,

by injecting warm water containing twenty to
forty drops of *Tr. Aconite* to the pint. By
repeating this application at every paroxysm,
patients recover rapidly, each succeed ng
attack being lighter, and the interval between being longer. until they cease entirely.
It may be used with much benefit in the
same manner. for *Hysteritis*, as well as recent
cases of *Leucorrhœa*. It is the most valuable
remedy applied to the *Eye* for a *wound* of that
organ. In *Gonorrhœa*, it is more valuable as
a local remedy, than most of those now in
use. It will frequently cure alone. In this
case, one part of tincture is to be used with
nine parts of warm water.

BELLADONNA has great power as a local
remedy in *Erysipelas*, to be applied with
water in proportion of ten drops of the *Tr.* to
a gill of warm water. It is also of much value
applied to the surface of inflamed breasts;
also injected when there is inflammation of
the *uterus*, with pressing pains as though the
bowels would be pressed out. *Very valuable*
in parturition when there is rigidity of the *os
uteri*, with fullness of the head and throbbing
of the temples. It has the specific power to
relax circular fibres without affecting the
longitudinal.

CALENDULA is applied to wounds, *incised*
and *lacerated*, promoting healing by the first
intention. It is a valuable application for
wounds in scrofulous persons, which tend to

suppurate rather than heal by the first intention. It is also useful in old sores. The
Calendula Cerate is one of the best dressings
for any abraded surface.

CONIUM is valuable as a *palliative* upon
cancerous tumors. As a *curative remedy* it is
useful in chronic ophthalmia, especially the
purulent of children; useful also for *indurat-
ed swellings*.

THUYA is a specific when locally used for
Sycosis, also for fungoid cancerous tumors. I
have cured well-marked cases of *Fungus
Hæmatodes* with the Tinct. Thuya applied to
the surface of the tumor.

The *Thuja Cerate* is a valuable application
for malignant ulcers

CORNUS-SERICEA will often cure malignant
ulcers both of the breast and uterus, used as
a wash.

ARSENICUM acts favorably on cancers, and
is a specific when applied to the surface of
carbuncle.

IPECAC. acts very beneficially when applied
to the surface where there is high fever, with
nausea and vomiting. Half an ounce of *Tr.*
Ipecac to two quarts of tepid water, applied
with a sponge to the whole surface, acts like
magic in yellow fever, allaying the nausea,
producing free and health-restoring perspiration.

RHUS-TOXICODENDRON applied with water at
the strength of thirty drops of the *Tr.* to a

gill, to parts affected with *Rheumatism*, acts very beneficially. It is also a most valuable application at half the above strength upon parts affected with Erysipelas, when the surface is swollen, and there are vesicles filled with fluid like a blister in burns. It is also useful for sores that exist as the chronic effects of burns when the proper treatment had not been used in the beginning, and the healing process was never perfected. *Rhus Cerate* is a very useful application to irritable ulcers.

Hepar-sulphur-kali is a specific for *Itch and Scald Head*, applied in form of a wash with twenty to thirty drops of *Tr. Hepar-sul.* to a gill of water. Also for ill-conditioned scrofulous ulcers, generally.

Cuprum-aceticum. — (*Acetate of Copper. Verdigris*) applied to *cancerous* ulcers of the face, *Lupus* or *Noli-me-tangere*, in the early stage, will in most cases effect a perfect cure, especially if for a week previously the part has been wet daily with *Tr. Thuja*. The best mode of applying the *acetate* is to mix the impalpable powder, as prepared for paint, with some substance to form a cerate, as equal parts of bees-wax and mutton suet, with one-fifteenth to one-hundredth part of the pure *acetate* as found in the bottom of the can, when prepared in oil for paint; heat all together and stir until cool. This forms a good plaster for covering and shielding the sore while its medicinal property is in the *Cuprum-*

aceticum diluted as above. It is quite useful for any ill-conditioned ulcer.

ACETIC-ACID is a most efficient remedy applied to old irritable *varicose ulcers* on the limbs of females who have suffered from *Phlegmasia Dolens*, (milk leg.) It may be applied as a wash to the part once or twice a day at the strength of 1-20th of the acid with water, or in the form of good cider vinegar. The manufactured vinegar of the cities does *not* usually contain acetic-acid.

ARUM-TRIPHYLLUM is a specific to allay the inflammation and excessive pain in *scrofulous swellings* of the neck, (*Kings Evil*.) The pure drug in powder, wet with warm water, or the green root bruised so as to form a poultice, is to be applied over the swelling. It soon discusses the swelling, or if pus has already formed, allays the pain, and brings the pus to the surface, and if continued, disposes it to heal rapidly.

BAPTISIA-TINCTORIA applied as a poultice either in the powdered drug, or with some other substance wet with the infusion or *Tr.*, *arrests gangrene* in a short time. It is especially useful for threatened or actual gangrene arising from *lacerated* wounds or scalds with wounds, as in accidents connected with the explosion of steam boilers; when we often have scalds and lacerations in the same wound.

HYDRASTIS-CANADENSIS used as a gargle in

a putrid state of the throat in malignant *Scarlet fever*, arrests the destructive process *at once.* It is also a most excellent application for inflamed eyes in the second or sub-acute stage, also as an injection in gonorrhœa and leucorrhœa, one part tincture to twenty parts water.

Prophylactics. — (*Preventives of D'sease.*) TO PREVENT SCARLET FEVER —Give Belladonna at the third attenuation, three to six pellets, according to the age of the child, every morning, during the prevalence of the epidemic. This is for the common or mild form of the disease. If the prevailing epidemic is of the *malignant* kind, producing fatal ulcerations of the throat, give *Bell.* once in two days and *Mercurius-corrosivus* at the 3d attenuation on the alternate day. While *Bell.* is a very certain preventive of the common eruptive Scarlatina, it is not as certain to prevent the *malignant* form. Though it renders the latter much more mild the *Mer.-cor.* is necessary to ward it off entirely, or so modify as to divest it of the dangerous features.

TO PREVENT YELLOW FEVER.—Take *Aconite, Belladonna* and *Macrotys,* 1st of each in rotation one dose a day. If there is any headache, or pains occur in other parts of the body, or a languid feeling, take a dose twice or three times a day in rotation.

TO PREVENT BILIOUS FEVER OR AGUE.—Take *Podophyllum, Baptisia* and *Gelseminum* 1st in rotation, one dose at night, and if symptoms of fever, as headache and loss of appetite, or bad taste in the mouth-in the morning appear, take a dose three times a day, and refrain entirely from food for one or two days.

To PREVENT TYPHOID FEVER. — When exposed as in nursing the sick, take *Baptisia* 2d, and *Macrotys* 2d, a dose three times a day.

To PREVENT SMALL-POX.—Use *Macro'ys* 1st night and morning, and if nursing or exposed frequently, use it every four hours.

To PREVENT CHOLERA. — *Camphor* (*pellets medicated* with the pure tincture) *Veratrum* 3d, and *Arsenicum* 3d, should be taken in rotation—a dose morning, noon and night, in the order named; so as to take a dose of each every twenty-four hours. If any sense of weakness or trembling comes on, use the *Camphor* oftener; if pain or uneasiness in the bowels threatening diarrhœa, use the *Veratrum*, and for increased thirst with uneasiness at the stomach *Arsenicum* more frequently.

To PREVENT DIARRHŒA.—Where it is prevailing as an *epidemic*, *Ipecac.* at night, and *Veratrum* in the morning will often suffice. For *teething children* give *Ipecac.* and *Chamomilla* in the same manner.

To PREVENT DYSENTERY.—In hot weather when bilious diseases prevail, use *Mercurius* 3d, *Podophyllum* 2d, and *Leptandrin* 1st in rotation, giving one dose a day. In the winter, or when *typhoid fevers* prevail, use *Mercurius* and *Rhus-tox.* alternately a dose every day.

To PREVENT ITCH.—A dose of *Sulphur,* or rubbing a little flour of sulphur on the hands, will generally suffice.

To PREVENT COLDS.—Keep the *arms, hands* and *chest* well clothed and warm. *Affecting* the *head* as *catarrh,* or the pelvic regions keep the *feet and ankles warm and dry.* Affecting joints and muscles as *Rheumatism*—protect the *spine* (back) from colds and currents of air. After an accidental exposure as by getting the feet wet, or being caught in a shower, drink *bountifully* of cold water, and take a dose of *Nux-vomica;* followed in an hour by Aconite if any chilliness is felt, or *Copaiva* if the head is stuffed up.

[**Treatment of Poisoning.** — From Dr. Hering's treatise on " Poisons," (Jahr's Clini. cal Guide.)

In treating a case of poisoning, two things are required : 1) *Removal of the exciting cause ;* and 2) *Treatment of the disease occasioned by the poisoning.*

The removal of the poisonous substances should be effected by the simplest and most innocent method, either with the finger, or, if this should be impossible, as in the case of poisons that had been swallowed, we recommend the following means suggested by *Hahnemann* and *Hering.*

1) Excite *vomiting* or *stool* by the simplest

means, copious administration of *tepid water*, *irritating the fauces* by means of a *feather* or something similar; placing on the tongue a pinch of *salt, snuff* or *mustard;* or, if neither of these means should be sufficient, resort to *injections of tobacco-smoke.*

2) *Neutralize the poison* by means of: The *white of an egg, vinegar,* or *lemon-juice, coffee, camphor, milk, oil, soap, mucilaginous drinks, tea, wine, sugar;* or, in some cases: *ammoniacal-gas, iron-rust, charcoal, kitchen-salt, epsom-salts, sweet almond oil, spiritus-nitr.-dulc., potash,* boiled *starch,* &c.

Particular indications:

WHITE OF AN EGG, dissolved in a sufficient quantity of water, and used as a drink, especially for: Metallic substances, such as, quicksilver, corrosive sublimate, verdigris, tin, lead, and sulphuric-acid; when the patient complains of violent pains in the stomach or abdomen, with tenesmus, or diarrhœa and pains at the anus.

VINEGAR: Antidotes poisoning with alkaline substances; but is hurtful in cases of poisoning with mineral acids, corrosive vegetable substances, Arsenic, and a large quantity of salts. In many cases it removes the ill effects of *Aconite, Opium, narcotic substances, poisonous mushrooms, belladonna, carbonic-acid gas, hepar-sulphuris, poisonous muscles* and *fish,* and even of *adipic acid.* The vinegar may be drank or administered by the rectum, alternately with mucilaginous substances. The vinegar should be as pure as possible. Crab-vinegar is, of itself, poisonous.

COFFEE: *Strong black coffee*, the beans being little roasted, and drank as hot as possible. Indispensable for a large number of poisons, especially when causing *drowsiness, intoxication, loss of consciousness,* or *mental derangement, delirium,* &c., in general, antidoting narcotic substances, such as: *Opium, nux-vom., belladonna, narcotic mushrooms, poisonous sumach, bitter almonds, prussic acid* and all those substances containing it, *Bell., colocynth, valer., cicuta* and *cham.* In case of poisoning with *Antimony, Phosphor.* and *Phosphoric acid,* coffee is no less indispensable.

CAMPHOR: Principal antidote of all vegetable substances, especially such as have a *corrosive* effect, or when *vomiting* and *diarrhœa, pale face, cold extremities* and *loss of consciousness* are present. Camphor is a specific remedy for the ill effects of poisoning insects, especially *cantharides,* whether administered internally or externally. Likewise for the effects of so-called, *worm-medicines, tobacco, bitter almonds,* and other fruits containing *prussic acid.* It is likewise useful for the secondary affections remaining after poisoning with *acids, salts, metals, phosphorus,* poisonous mushrooms, &c., after the poisonous substance itself had been removed from the stomach by means of vomiting, &c.

MILK: Less useful than is supposed. To procure an artificial covering or envelop for the poison, mucilaginous substances are to be preferred. *Fat milk* (or *cream*) is suitable in all cases where *oil* is, and hurtful where oil is. Curdled or sour milk is suitable or not suitable in all cases where vinegar is or is not.

OLIVE OIL: Less useful than is believed. It is of no use in cases of metalic poisoning, and even hurtful in cases of poisoning with Arsenic. It is very bad for the ill effects of *Canthar.* This remark applies to poisoning with any other insect, or if the poison should have got into one's eye. Oil may be used to facilitate the extraction of insects from the ear in case they should have got into it. Oil is most suitable for poisoning with corrosive acids, such as: *nitric acid, sulphuric acid,* &c. It is sometimes useful in cases of poisoning with alkalies, to be administered alternately with vinegar, and in cases of poisoning with mushrooms.

MUCILAGINOUS SUBSTANCES, drinks or injections of mucilaginous substances, should be resorted to in cases of poisoning with alkalies, especially when administered alternately with vinegar.

SOAP, *common castile soap*, dissolved in four times its bulk of hot water and drank, is one of the best remedies in many cases of poisoning. It may be drank by the cupful,—a cupful every two, three, or four minutes, in all cases where the white of an egg is indicated but does not produce sufficient relief. Soap is particularly useful in all cases of poisoning with metallic substances, especially *Arsenic, lead*, &c. Likewise for poisoning with corrosive acids, such as : *Sulphuric acid, nitric acid, &c.,* with *alum, corrosive vegetable substances, castor oil,* &c. Soap is hurtful in cases of poisoning with alkalies, such as : *Lye, nitrate of silver, potash, soda, oleum tartari, ammonium muriaticum,* (Salmiac) *ammonium carbonicum, caustic* or *burnt lime, barytes,* &c.

SUGAR, or sugar water, one of the best remedies in many cases. In case of poisoning with mineral acids or *alkalies,* it is best to resort at once to the specific antidote, though sugar is not hurtful. In cases of poisoning with *metallic substances,* various kinds of *paint, verdigris, copper, sulphate of copper, alum,* &c., sugar is preferable to every other remedy, and not till the patient has been relieved by the sugar, administer the *white of an egg* or soap-water alternately with sugar. Sugar is likewise an excellent antidote in cases of poisoning with *arsenic,* or *corrosive vegetable substances.*

Of the other antidotes, use :

AMMONIACAL GAS: For *alcohol, bitter almonds, prussic acid.*

IRON-RUST : For *Arsenic.*

EPSOM-SALTS : For *alkaline poisons.*

CHARCOAL : For *foul fish, foul meat, poisonous mushrooms, poisonous muscles, &c.*

KITCHEN-SALT: For *nitrate of silver* and *poisonous wounds.*

MAGNESIA: For *acids.*

SWEET-ALMOND OIL: For *acids.*

POTASH: For *acids.*

STARCH: For *iodine.*

SPIRITS OF NITRE: For *alkaline poisons* and *animal substances.*

TEA: For *adipic acid* and *poisonous honey.*

WINE: For *noxious vapors* and *poisonous mushrooms.*

The first thing we have to do, in treating a case of poisoning, is to remove the poison by vomiting, and then to administer suitable antidotes.

If we should not be able to ascertain what kind of poison had been swallowed, we should first administer the white of an egg, and, if there should be stupefaction, *Coffee.* If we should know that the poison is:

a) A metallic substance, we have to give: first the *white off an egg, sugar-water, soap-water*, and afterwards, for the remaining ailments: *Sulph.*, which is a real antidote to metals.

b) If *acids* and *corrosive substances*, give: 1) *Soap-water;* 2) *Magnesia* dissolved in water; 3) *Chalk water;* 4) *Alkalie* or *potash* dissolved in water, taking a tablespoonful as long as the vomiting continues. Afterwards mucilaginous drinks, and alternately *Coff.* and *Op.* as homœopathic antidotes.

As regards the remaining ailments, give

Puls. for sulphuric acid; *Bry.* for muriatic acid; *Acon.* for the other acids, and especially crab-apple vinegar.   If the skin should have been corroded by poisons, apply soap-water, or a watery solution of *Caust.;* and if corrosive substances should have got into the eyes, apply *sweet almond-oil,* or *fresh unsalted butter.*

c) For *alkaline substances:* 1) *Vinegar and water* in large quantities; 2) *Lemon-juice,* or acids from other fruits, diluted with much water; 3) *Sour milk;* 4) *Mucilaginous drinks,* or injections.   *Vinegar* is hurtful in cases of poisoning with Barytes; but epsom-salt dissolved in water, renders good service; afterwards, *Camph.* or *Nitr.-spir.*   The secondary effects of poisoning with potash, require: *Coff.* or *Carb-v.;* and those of poisoning with spirits of Ammonia, *Hep.*

d) For the inhalation of *noxious vapors:* Sprinkle the patient with *vinegar and water,* or let him inhale the *vapors of a solution of chlore;* afterwards, after the return of consciousness, give *black coffee,* or a few doses of *Op.* or *Bell.*

e) For *vegetable poisons:* 1) *Camphor,* by olfaction, or sometimes a drop of the spirits of camphor on sugar: 2) *Black coffee* or *vinegar,* especially for narcotic vegetable juices. The best antidotes for corrosive vegetable juices, are soap-water and milk.

f) For *animal poisons:* For *toad-poison*, or similar poisons, if they should have got into the stomach, give powdered charcoal, stirred up with oil or milk; or let the patient smell of the sweet spirits of nitre, if bad symtoms should set in, and afterwards give *Ars.*—If a poison of this kind should have got into the eye, give *Acon.*

As regards the *wounds* or bites inflicted by poisonous animals, *Hering* proposes the following mode of treatment: For the *bites* of *poisonous serpents*, *mad dogs*, or other poisonous animals, apply *heat at a distance*, for which purpose any thing may be used which is handy at the time: a red-hot iron, incandescent piece of coal, or even a burning cigar; hold this as near as possible without burning the skin. The heat should be kept up uniformly, and should be concentrated upon the wound exclusively. The edges of the wound should be covered over with *oil* or *fat*, and this should be repeated as often as the skin gets dry. If no oil or fat can be had, use *soap*, or even saliva. Wipe off carefully every thing which is discharged from the wound. Continue the application of heat until the patient feels chilly and stretches himself; if this should take place too speedily, continue to apply the heat for about an hour, or until the effects of the poison commence to disappear.

At the same time administer internal remedies. In the case of a serpent's bite, give the patient a swallow of salt-water from time to time, or a pinch of salt or powder, or a few pieces of garlic.

If, nevertheless, dangerous symptoms should set in, give a tablespoonful of wine or brandy every 2 or 3 minutes; continue this until the symptoms abate, and repeat the brandy at every return of a paroxysm.

If the stitching pains should increase in violence, and be felt nearer the heart; if the wound, at the same time, should be bluish, checkered like marble and swollen, with vomiting, vertigo and diarrhœa, give *Ars.* 30, and another dose in half an hour, if the symptoms should continue to get worse, or only in 3 hours, if they should remain unchanged; if an improvement should set in after the first dose, do not repeat the medicine until the symptoms get worse again.

If *Ars.*, even if repeated, should have no effect, give *Bell.* In some cases *Senega* may be tried. The chronic sequelæ of the bite of a serpent require: *Phos.-ac.* and *Merc.*

If the bite should have been inflicted by a mad dog, apply *heat at a distance* as above.

Wounds which have become poisonous in consequence of decayed animal matter or pús having got into them, require *Ars.*

To guard against unpleasant consequences in case we should have to touch decayed animal substances, poisonous wounds or ulcers, or men and animals infected with contagious diseases, we should hold our hands for ten or fifteen minutes near as strong a heat as can be borne, and afterwards wash them with soap. The use of *Chlore* in such cases is well known.]

Permanganate of potash is highly useful in such cases. Use a solution in water 1 to 9.

[Dr. Marshall Hall's Instructions for Restoring Persons Apparently Drowned.

Treat the parties *instantly, on the spot, in the open air*, exposing the face and chest to the breeze, (except in severe weather.)

.1) *To Clear the Throat*—Place the patient gently on the face, and one wrist (of the patient) under the forehead. All fluids and the tongue itself then fall forwards, leaving the entrance to the windpipe free. If there be breathing, wait and watch; if not, or if it fail,

2) *To Excite Respiration*—Turn the patient well and instantly on his side, and excite the nostrils with snuff, or the throat with a feather, &c., and dash cold water on the face, previously rubbed warm. If there be no success, *lose not a moment*, but instantly

3) *To Imitate Respiration* — Replace the patient on his face, raising and supporting the chest well on a folded coat, or other article of dress; turn the body very gently on the side and a little beyond, and then briskly on the face, alternately; repeating these measures deliberately, efficiently and perseveringly, fifteen times in the minute, occasionally varying

the side. (When the patient reposes on the chest, this cavity is compressed by the weight of the body, and expiration takes place; when he is turned on the side, this pressure is removed and inspiration occurs.) When the prone position is resumed, make equable but efficient pressure, with brisk movement along the back of the chest; removing it immediately before rotation on the side. (The first measure augments the expiration, the second commences inspiration.)

The result is inspiration—and *if not too late* —life.

4. *To induce Circulation and Warmth—* Meantime rub the limbs upwards, with a firm grasping pressure and with energy, using handkerchiefs, &c. By this measure the blood is propelled along the veins towards the heart. Let the limbs be thus warmed and dried and then clothed, the bystanders supplying the requisite garments, (as one a coat, another pantaloons, &c., from their own persons, if necessary.) Avoid the continuous warm bath and the position on or inclined to the back.]

INDEX.